D1196870

The Soil and Health

Culture of the Land

A Series in the New Agrarianism

This series is devoted to the exploration and articulation of a new agrarianism that considers the health of habitats and human communities together. It is intended to demonstrate how agrarian insights and responsibilities can be worked out in diverse fields of learning and living: history, science, art, politics, economics, literature, philosophy, religion, urban planning, education, and public policy. Agrarianism is a comprehensive worldview that appreciates the intimate and practical connections which exist between humans and the earth. It stands as our most promising alternative to the unsustainable and destructive ways of current global, industrial, and consumer culture.

Series Editor

Norman Wirzba, Georgetown College, Kentucky

Advisory Board

Wendell Berry, Port Royal, Kentucky

Ellen Davis, Duke University, North Carolina

Patrick Holden, Soil Association, United Kingdom

Wes Jackson, Land Institute, Kansas

Gene Logsdon, Upper Sandusky, Ohio

Bill McKibben, Middlebury College, Vermont

David Orr, Oberlin College, Ohio

Michael Pollan, University of California at Berkeley, California

Jennifer Sahn, *Orion* magazine, Massachusetts

Vandana Shiva, Research Foundation for Science,
Technology and Ecology, India

William Vitek, Clarkson University, New York

The

Soil
and
Health

A Study of
Organic Agriculture

SIR ALBERT HOWARD

With a New Introduction
by Wendell Berry

THE UNIVERSITY PRESS OF KENTUCKY

Publication of this volume was made possible in part by a grant
from the National Endowment for the Humanities.

Published in 2006 by The University Press of Kentucky
Scholarly publisher for the Commonwealth,
serving Bellarmine University, Berea College, Centre College of Kentucky,
Eastern Kentucky University, The Filson Historical Society, Georgetown
College, Kentucky Historical Society, Kentucky State University, Morehead
State University, Murray State University, Northern Kentucky University,
Transylvania University, University of Kentucky, University of Louisville,
and Western Kentucky University.
All rights reserved.

Editorial and Sales Offices: The University Press of Kentucky
663 South Limestone Street, Lexington, Kentucky 40508-4008
www.kentuckypress.com

Library of Congress Cataloging-in-Publication Data

Howard, Albert, Sir, 1873–1947.
    The soil and health : a study of organic agriculture / Sir Albert Howard.
        p. cm. — (Culture of the land: a series in the new agrarianism)
    Originally published in 1947 by The Devin-Adair Company.
    Includes bibliographical references and index.
    ISBN-13: 978-0-8131-9171-3 (pbk. : alk. paper)
    ISBN-10: 0-8131-9171-8 (pbk. : alk. paper)
    1. Organic farming. 2. Organic gardening. 3. Plant diseases.
    I. Title. II. Series.
    S605.5.H67 2007
    631.5'84—dc22   2006025168

This book is printed on acid-free recycled paper meeting
the requirements of the American National Standard
for Permanence in Paper for Printed Library Materials.

Manufactured in the United States of America.

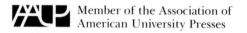 Member of the Association of
American University Presses

"The civilized nations—Greece, Rome, England—
have been sustained by the primitive forests which
anciently rotted where they stood. They survive as
long as the soil is not exhausted."
                    —Thoreau, *Walking and the Wild*

"The staple foods may not contain the same
nutritive substances as in former times. . . .
Chemical fertilizers, by increasing the abundance
of the crops without replacing all the exhausted
elements of the soil, have indirectly contributed to
change the nutritive value of cereal grains and of
vegetables. . . . Hygienists have not paid sufficient
attention to the genesis of diseases. Their studies
of conditions of life and diet, and of their effects
on the physiological and mental state of modern
man, are superficial, incomplete, and of too short
duration. They have, thus, contributed to the
weakening of our body and our soul."
                    —Alexis Carrel, *Man the Unknown*

"The preservation of fertility is the first duty of
all that live by the land. . . . There is only one rule
of good husbandry—leave the land far better than
you found it."
                    —George Henderson, *The Farming Ladder*

CONTENTS

ILLUSTRATIONS

PLATES

FIGURES

NEW INTRODUCTION

In 1964 my wife, Tanya, and I bought a rough and neglected little farm on which we intended to grow as much of our own food as we could. My editor at the time was Dan Wickenden, who was an organic gardener and whose father, Leonard Wickenden, had written a practical and inspiring book, *Gardening with Nature,* which I bought and read. Tanya and I wanted to raise our own food because we liked the idea of being independent to that extent, and because we did not like the toxicity, expensiveness, and wastefulness of "modern" food production. *Gardening with Nature* was written for people like us, and it helped us to see that what we wanted to do was possible. I asked Dan where his father's ideas had come from, and he gave me the name of Sir Albert Howard. My reading of Howard, which began at that time, has never stopped, for I have returned again and again to his work and his thought. I have been aware of his influence in virtually everything I have done, and I don't expect to graduate from it. That is because his way of dealing with the subject of agriculture is also a way of dealing with the subject of life in this world. His thought is systematic, coherent, and inexhaustible.

Sir Albert Howard was born in 1873 to a farming family in Shropshire, and he died in 1947. He published several books and also many articles in agricultural journals. His best-known books, *An Agricultural Testament* (1940) and *The Soil and Health* (1947), were addressed both to general readers and to his fellow scientists.

An Agricultural Testament and *The Soil and Health* are products of Howard's many years as a government scientist in India, during which he conceived, and set upon sound scientific footing, the kind of agriculture to which his followers have applied the term "organic." But by 1940, when the first of these books was published, the industrialization of agriculture had already begun. By 1947, when *The Soil and Health* was published, World War II had proved the effectiveness of the mechanical and chemical technology that in the coming decades would radically alter both the practice of agriculture and its underlying assumptions.

This "revolution" marginalized Howard's work and the kind of agriculture he advocated. So-called organic agriculture survived only on the margin. It was practiced by some farmers of admirable independence and good sense and also by some authentic nuts. In the hands of the better practitioners, it was proven to be a healthful, productive, and economical way of farming. But while millions of their clients spent themselves into bankruptcy on industrial supplies, the evangelists of industrial agriculture in government and the universities ignored the example of the successful organic farmers, just as they ignored the equally successful example of Amish farming.

Meanwhile, Howard's thought, as manifested by the "organic movement," was seriously oversimplified. As it was understood and prescribed, organic agriculture improved the health of crops by building humus in the soil, and it abstained from the use of toxic chemicals. There is nothing objectionable about this kind of agriculture, so far as it goes, but it does not go far enough. It does not conceive of farms in terms of their biological and economic structure, because it does not connect farming with its ecological and social contexts. Under the current and now official definition of organic farming, it is possible to have a huge "organic" farm that grows only one or two crops, has no animals or pastures, is entirely dependent on industrial technology and economics, and imports all its fertility and energy. It was precisely this sort of specialization and oversimplification that Sir Albert Howard worked and wrote against all his life.

At present this movement (if we can still apply that term to an effort that is many branched, multicentered, and always in flux) in at least some of its manifestations appears to be working decisively against such oversimplification and the industrial gigantism that oversimplification allows. Some food companies as well as some consumers now understand that only the smaller family farms, such as those of the Amish, permit the diversity and the careful attention that Howard's standards require.

Howard's fundamental assumption was that the processes of agriculture, if they are to endure, have to be analogous to the processes of nature. If one is farming in a place previously forested, then the farm must be a systematic analogue of the forest, and the farmer must be a student of the forest. Howard stated his premise as an allegory:

> The main characteristic of Nature's farming can . . . be summed up in a few words. Mother earth never attempts to farm without live stock; she always raises mixed crops; great pains are taken to preserve the soil and to prevent erosion; the mixed vegetable and animal wastes are converted into humus; there is no waste; the processes of growth and the processes of decay balance one another; ample provision is made to maintain large reserves of fertility; the greatest care is taken to store the rainfall; both plants and animals are left to protect themselves against disease.[1]

Nature is the ultimate circumstance of the practical or economic world. We cannot escape either it or our dependence on it. It is, so to speak, its own context, whereas the context of agriculture is, first, nature and then the human economy. Harmony between agriculture and its natural and human contexts would be health, and health was the invariable standard of Howard's work. His aim always was to treat "the whole problem of health in soil, plant, animal, and man as one great subject."[2] And Louise Howard spells this out in *Sir Albert Howard in India:*

> A fertile soil, that is, a soil teeming with healthy life in the shape of abundant microflora and microfauna, will bear healthy plants, and these, when consumed by animals and man, will confer health on animals and man. But an infertile soil, that is, one lacking sufficient microbial, fungous, and other life, will pass on some form of deficiency to the plant, and such plant, in turn, will pass on some form of deficiency to animal and man.[3]

This was Howard's "master idea," and he understood that it implied a long-term research agenda, calling for "a boldly revised point of view and entirely fresh investigations."[4]

His premise, then, was that the human economy, which is inescapably a land-using economy, must be constructed as an analogue of the organic world, which is inescapably its practical context. And so he was fundamentally at odds with the industrial economy, which sees creatures, including humans, as machines, and agriculture, like ultimately the entire human economy, as an analogue of an industrial system. This was, and is, the inevitable and characteristic product of the dead-end materialism that is the premise both of industrialism and the science that supports it.

Howard understood that such reductionism could not work for agriculture:

> But the growing of crops and the raising of live stock belong to biology, a domain where everything is alive and which is poles asunder from chemistry and physics. Many of the things that matter on the land, such as soil fertility, tilth, soil management, the quality of produce, the bloom and health of animals, the general management of live stock, the working relations between master and man, the *esprit de corps* of the farm as a whole, cannot be weighed or measured. Nevertheless their presence is everything: their absence spells failure.[5]

This understanding has a scientific basis, as it should have, for Howard was an able and conscientious scientist. But I think it comes also from intuition, and probably could not have come otherwise. Howard's intuition was that of a man who was a farmer by birth and heritage and who was a sympathetic as well as a scientific observer of the lives of plants, animals, and farmers.

If the farm is to last—if it is to be "sustainable," as we now say—then it must waste nothing. It must obey in all its processes what Howard called "the law of return." Under this law, agriculture produces no waste; what is taken from the soil is returned to it. Growth must be balanced by decay: "In this breaking down of organic matter we see in operation the reverse of the building-up process which takes place in the leaf."[6]

The balance between growth and decay is the sole principle of stability in nature and in agriculture. And this balance is never static, never finally achieved, for it is dependent upon a cycle, which in nature, and within the limits of nature, is self-sustaining, but which in agriculture must be made continuous by purpose and by correct methods. "This cycle," Howard wrote, "is constituted of the successive and repeated processes of birth, growth, maturity, death, and decay."[7]

The interaction, the interdependence, of life and death, which in nature is the source of an inexhaustible fecundity, is the basis of a set of analogies, to which agriculture and the rest of the human economy must conform in order to endure, and which is ultimately religious, as Howard knew: "An eastern religion calls this cycle the Wheel of Life . . . Death supersedes life and life rises again from what is dead and decayed."[8]

The maintenance of this cycle is the practical basis of good farming and its moral basis as well:

> [T]he correct relation between the processes of growth and the processes of decay is the first principle of successful farming. Agriculture must always be balanced. If we speed up growth we must accelerate decay. If, on the other hand, the soil's reserves are squandered, crop production ceases to be good farming: it becomes something very different. The farmer is transformed into a bandit.[9]

It seems to me that Howard's originating force, innate in his character and refined in his work, was his sense of context. This made him eminent and effective in his own day, and it makes his work urgently relevant to our own. He lacked completely the specialist impulse, so prominent among the scientists and intellectuals of the present-day university, to see things in isolation.

He himself began as a specialist, a mycologist, but he soon saw that this made him "a laboratory hermit," and he felt that this was fundamentally wrong:

> I was an investigator of plant diseases, but I had myself no crops on which I could try out the remedies I advocated: I could not take my own advice before offering it

xvii

to other people. It was borne in on me that there was a wide chasm between science in the laboratory and practice in the field, and I began to suspect that unless this gap could be bridged no real progress could be made in the control of plant diseases: research and practice would remain apart: mycological work threatened to degenerate into little more than a convenient agency by which—provided I issued a sufficient supply of learned reports fortified by a judicious mixture of scientific jargon—practical difficulties could be side-tracked.[10]

The theme of his life's work was his effort to bridge this gap. The way to do it was simply to refuse to see anything in isolation. Everything, as he saw it, existed within a context, outside of which it was unintelligible. Moreover, every problem existed within a context, outside of which it was unsolvable. Agriculture, thus, cannot be understood and its problems cannot be solved without respect to context. The same applied even to an individual plant or crop. And this respect for context properly set the standard and determined the methodology of agricultural science:

> The basis of research was obviously to be investigation directed to the whole existence of a selected crop, namely, 'the plant itself in relation to the soil in which it grows, to the conditions of village agriculture under which it is cultivated, and with reference to the economic uses of the product'; in other words research was to be integral, never fragmented.[11]

If nothing exists in isolation, then all problems are circumstantial; no problem resides, or can be solved, in anybody's department. A disease was, thus, a symptom of a larger disorder. The following passage shows as well as any the way his mind worked:

> I found when I took up land in India and learned what the people of the country know, that the diseases of plants and animals were very useful agents for keeping me in order, and for teaching me agriculture. I have learnt more from the diseases of plants and animals than I have

from all the professors of Cambridge, Rothamsted and other places who gave me my preliminary training. I argued the matter in this way. If diseases attacked my crops, it was because I was doing something wrong. I therefore used diseases to teach me. In this way I really learnt agriculture—from my father and from my relatives and from the professors I only obtained a mass of preliminary information. Diseases taught me to understand agriculture. I think if we used diseases more instead of running to sprays and killing off pests, and if we let diseases rip and then found out what is wrong and then tried to put it right, we should get much deeper into agricultural problems than we shall do by calling in all these artificial aids. After all, the destruction of a pest is the evasion of, rather than the solution of, all agricultural problems.[12]

The implied approach to the problem of disease is illustrated by the way Howard and his first wife, Gabrielle, dealt with the problem of indigo wilt:

In fifteen years £54,207 had been spent on research, at that time a large sum. Yet the Imperial Entomologist could find no insect, the Imperial Mycologist no fungus, and the Imperial Bacteriologist no virus to account for the plague.

The Howards proceeded differently. Their start was to grow the crop on a field scale and in the best possible way, taking note of local methods. Their observation was directed to the whole plant, above and below ground; they followed the crop throughout its life history; they looked at all the surrounding circumstances, soil, moisture, temperature. But they looked for no virus, no fungus, and no insect.[13]

And it was the Howards who solved the problem. The plants were wilting, they found, primarily because the soils were becoming waterlogged during the monsoon, killing the roots; the plants were wilting and dying from starvation. It was a problem of management, and it was solved by changes in management. But it could not have been solved except by studying the whole plant in its whole context.

Because he refused to accept the academic fragmentation that had become conventional by his time, Howard, of course, was "accused of invading fields not his own,"[14] and this he had done intentionally and in accordance with "the guiding principle of the closest contact between research and those to be served."[15]

Agriculture is practiced inescapably in a context, and its context must not be specialized or simplified. Its context, first of all, is the nature of the place in which it is practiced, but it is also the society and the economy of those who practice it. And just as there are penalties for ignoring the natural context, so there are penalties for ignoring the human one. As Howard saw it, the agricultural industrialists' apparent belief that food production could be harmlessly divorced from the economic interest of farmers needlessly repeats a historical failure:

> Judged by the ordinary standards of achievement the agricultural history of the Roman Empire ended in failure due to inability to realize the fundamental principle that the maintenance of soil fertility coupled with the legitimate claims of the agricultural population should never have been allowed to come in conflict with the operations of the capitalist. The most important possession of a country is its population. If this is maintained in health and vigour everything else will follow; if this is allowed to decline nothing, not even great riches, can save the country from eventual ruin.[16]

The obligation of a country's agriculture, then, is to maintain its people in health, and this applies equally to the people who eat and to the people who produce the food.

Howard accepted this obligation unconditionally as the obligation also of his own work. He realized, moreover, that this obligation imposed strict limits both upon the work of farmers and upon his work as a scientist: first, neither farming nor experimentation should usurp the tolerances or violate the nature of the place where the work is done; and second, the work must respect and preserve the livelihoods of the local community. Before going to work, agricultural scientists are obliged to know both the place where their work is to be done and the people for whom they are working. It is remarkable that Howard came quietly, by thought and work, to these realizations a half century and more before they were forced upon us by the ecological and economic failures of industrial agriculture.

In India he used his training as a scientist and his ability to observe and think for himself, just as he would have been expected to do. But he also learned from the peasant farmers of the country, whom he respected as his "professors." He valued them for their knowledge of the land, for their industry, and for their "accuracy of eye."[17] He accepted also the economic and technological circumstances of those farmers as the limit within which he himself should do his work. He saw that it would be possible to ruin his clients by thoughtless or careless innovation:

> Often improvements are possible but they are not economic. . . . In India the cultivators are mostly in debt and the holdings are small. Any capital required for developments has to be borrowed. A large number of possible improvements are barred by the fact that the extra return is not large enough to pay the high interest on the capital involved and also to yield a profit to the cultivator.[18]

The reader may wish to contrast this way of thinking with that of the Green Revolution or with that of the headlong industrialization of American agriculture since World War II, in both of which the only recognized limit was technological, and in neither of which was there any concern for the ability of farmers or their communities to bear the costs.

Howard's solution to the problem was simply to do his work within the technological limits of the local farmers:

> The existing system could not be radically changed, but it might be developed in useful ways. This must never exceed what the cultivator could afford, and, in a way, also what he was used to. This principle Sir Albert kept in mind to the very end . . . his standard seems to have

xxi

been the possession of a yoke of oxen; when more power was needed, the presumption was that the second yoke could be borrowed from a neighbor. Thus the maximum draught contemplated was four animals.[19]

By the observance of such limits, Howard was enfolded consciously and conscientiously within the natural and human communities that he endeavored to serve.

No university that I have heard of, land-grant or other, has yet attempted to establish its curriculum and its intellectual structure on Sir Albert Howard's "one great subject," or on his determination to serve respectfully and humbly the local population. But a university most certainly could do so, and in doing so it could bring to bear all its disciplines and departments. In doing so, that is to say, it could become in truth a uni-versity.

At present, our universities are not simply growing and expanding, according to the principle of "growth" that is universal in industrial societies, but they are at the same time disintegrating. They are a hodgepodge of unrelated parts. There is no unifying aim and no common critical standard that can serve equally well all the diverse parts or departments.

The fashion now is to think of universities as industries or businesses. University presidents, evidently thinking of themselves as CEOs, talk of "business plans" and "return on investment," as if the industrial economy could provide an aim and a critical standard appropriate either to education or to research.

But this is not possible. No economy, industrial or otherwise, can supply an appropriate aim or standard. Any economy must be either true or false to the world and to our life in it. If it is to be true, then it must be *made* true, according to a standard that is not economic.

To regard the economy as an end or as the measure of success is merely to reduce students, teachers, researchers, and all they know or learn, to merchandise. It reduces knowledge to "property" and education to training for the "job market."

If, on the contrary, Howard was right in his belief that health is the "one great subject," then a unifying aim and a common critical standard are clearly implied. Health is at once quantitative and qualitative; it requires both sufficiency and goodness. It is comprehensive (it is synonymous with "wholeness"), for it must leave nothing out. And it is uncompromisingly local and particular; it has to do with the sustenance of particular places, creatures, human bodies, and human minds.

If a university began to assume responsibility for the health of its place and its local constituents, then all of its departments would have a common aim, and they would have to judge their place and themselves and one another by a common standard. They would need one another's knowledge. They would have to communicate with one another; the diversity of specialists would have to speak to one another in a common language. And here again, Howard is exemplary, for he wrote, and presumably spoke, a plain, vigorous, forthright English—no jargon, no condescension, no ostentation, no fooling around.

Wendell Berry
Port Royal, Kentucky

NOTES

The new introduction by Wendell Berry first appeared in *The Hudson Review*, vol. 59, no. 2 (Summer 2006).

1. Sir Albert Howard, *An Agricultural Testament* (Oxford: Oxford University Press, 1956), 4.

2. Sir Albert Howard, *The Soil and Health* (Lexington, Ky.: University Press of Kentucky, 2006), 11.

3. Louise E. Howard, *Sir Albert Howard in India* (Emmaus, Pa.: Rodale Press, 1954), 162.

4. Howard, *Soil and Health*, 11.

5. Howard, *Agricultural Testament*, 196.

6. Howard, *Soil and Health*, 22.

7. Ibid., 18.

8. Ibid.

9. Howard, *Agricultural Testament*, 25.

10. Howard, *Soil and Health*, 1–2.

11. Howard, *Sir Albert Howard in India*, 42.

12. Sir Albert Howard, as quoted in Howard, *Sir Albert Howard in India*, 190.

13. Howard, *Sir Albert Howard in India*, 170.

14. Ibid., 42.

15. Ibid., 44.

16. Howard, *Agricultural Testament*, 9.

17. Howard, *Sir Albert Howard in India*, 222, 228.

18. Sir Albert Howard, as quoted in Howard, *Sir Albert Howard in India*, 37–38.

19. Howard, *Sir Albert Howard in India*, 224.

PREFACE

The Earth's green carpet is the sole source of the food consumed by livestock and mankind. It also furnishes many of the raw materials needed by our factories. The consequence of abusing one of our greatest possessions is disease. This is the punishment meted out by Mother Earth for adopting methods of agriculture which are not in accordance with Nature's law of return. We can begin to reverse this adverse verdict and transform disease into health by the proper use of the green carpet—by the faithful return to the soil of all available vegetable, animal, and human wastes.

The purpose of this book is threefold: to emphasize the importance of solar energy and the vegetable kingdom in human affairs; to record my own observations and reflections, which have accumulated during some forty-five years, on the occurrence and prevention of disease; to establish the thesis that most of this disease can be traced to an impoverished soil, which then leads to imperfectly synthesized protein in the green leaf and finally to the breakdown of those protective arrangements which Nature has designed for us.

During the course of the campaign for the reform of agriculture, now in active progress all over the world, I have not hesitated to question the soundness of present-day agricultural teaching and research—due to failure to realize that the problems of the farm and garden are biological rather than chemical. It follows, therefore, that the foundations on which the artificial manure and poison spray industries are based are also unsound. As a result of this onslaught, what has been described as the war in the soil has broken out in many countries and continues to spread. The first of the great battles now being fought began in South Africa some ten years ago and has ended in a clear-cut victory for organic farming. In New Zealand the struggle closely follows the course of the South African conflict. The contest in Great Britain and the United States of America has only now emerged from the initial phase of reconnaissance, in the course of which the manifold weaknesses of the fortress to be stormed have been discovered and laid bare.

I am indebted to some hundreds of correspondents all over the world for sending me reports of the observations, experiments, and results which have followed the faithful adoption of Nature's great law of return. Some of this information is embodied and acknowledged in the pages of this book. A great deal still remains to be summarized and reduced to order—a labour which I hope soon to begin. When it is completed, a vast mass of material will be available which will confirm and extend what is to be found in these pages. Meanwhile a portion of this evidence is being recorded by Dr. Lionel J. Picton, O.B.E., in the *News-Letter on Compost* issued three times a year by the County Palatine of Chester Local Medical and Panel Committees at Holmes Chapel, Cheshire. By this means the story begun in their Medical Testament of 1939 is being continued and the pioneers of organic farming and gardening are kept in touch with events.

The fourth chapter on "The Maintenance of Soil Fertility in Great Britain" is very largely based on the labours of a friend and former colleague, the late Mr. George Clarke, C.I.E., who, a few days before his untimely death in May 1944, sent me the results of his study of the various authorities on the Saxon Conquest, the evolution of the manor, the changes it underwent as the result of the Domesday Book, and the enthronement of the feudal system till the decay of the open-field system and its replacement by enclosure.

The spectacular progress in organic farming and gardening which has taken place in South Africa and Rhodesia during the last few years owes much to the work of Captain Moubray, Mr. J. P. J. van Vuren, and Mr. G. C. Dymond, who have very generously placed their results at my disposal. Captain Moubray and Mr. van Vuren have contributed two valuable appendices, while Mr. Dymond's pioneering work on virus disease in the cane and on composting at the Springfield Sugar Estate in Natal has been embodied in the text. For the details relating to the breakdown of the cacao industry in Trinidad and on the Gold Coast and for a number of other suggestions on African and West Indian agriculture I am indebted to Dr. H. Martin Leake, formerly Principal of the Imperial College of Tropical Agriculture, Trinidad.

I have been kept in constant touch with the progress of organic farming and gardening in the United States of America by Mr. J. I. Rodale of Emmaus, Pa., the editor of *Organic Gardening*, who has started a movement in the New World which promises soon to become an avalanche. Mr. Rodale was the prime mover in bringing out the first American edition of *An Agricultural Testament* and is responsible for the simultaneous publication of this present book in the United States and of a special American issue of Lady Eve Balfour's stimulating work—*The Living Soil*.

In India I have made full use of the experience of Colonel Sir Edward Hearle Cole, C.B., C.M.G., on the Coleyana Estate in the Punjab, and of Mr. E. F. Watson's work on the composting of water hyacinth at Barrackpore. Messrs. Walter Duncan & Company have generously permitted Mr. J. C. Watson to contribute an appendix on the remarkable results he has obtained on the Gandrapara Tea Estate in North Bengal. In this fine property India and the rest of the Empire possess a perfect example of the way Nature's law of return should be obeyed and of what freshly prepared humus by itself can achieve.

I owe much to a number of the active members of the New Zealand Compost Club, and in particular to its former Honorary Secretary, Mr. T. W. M. Ashby, who have kept me fully informed of the results obtained by this vigorous association. The nutritional results obtained by Dr. G. B. Chapman, the President, at the Mount Albert Grammar School, which show how profoundly the fresh produce of fertile soil influences the health of schoolboys, have been of the greatest use. In Eire the Rev. C. W. Sowby, Warden of the College of St. Columba, Rathfarnham, Co. Dublin, and the Rev. W. S. Airy, Head Master of St. Martin's School, Sidmouth, have placed at my disposal the results of similar work at their respective schools. These pioneering efforts are certain to be copied and to be developed far and wide. Similar ideas are now being applied to factory canteen meals in Great Britain with great success, as will be evident from what Mr. George Wood has already accomplished at the Co-operative Wholesale Society's bacon factory at Winsford in Cheshire.

For furnishing full details of a large-scale example of successful mechanized organic farming in this country and of the great possibilities of our almost unused downlands I owe much to Mr. Friend Sykes. The story of Chantry, where the results of humus without any help from artificial manures are written on the land itself, provides a fitting conclusion to this volume.

In the heavy task of getting this book into its final shape I owe much to the care and devotion of my private secretary, Miss Ellinor Kirkham.

A. H.
14 Liskeard Gardens
Blackheath
London, S.E. 3

I

INTRODUCTION

AN ADVENTURE IN RESEARCH

M<small>Y</small> F<small>IRST</small> post was a somewhat unusual one. It included the conventional investigation of plant diseases, but combined these duties with work on general agriculture; officially I was described as Mycologist and Agricultural Lecturer to the Imperial Department of Agriculture for the West Indies.

The headquarters of the department were at Barbados. While I was here provided with a laboratory for investigating the fungous diseases of crops (mycology) and was given special facilities for the study of the sugar-cane, my main work in the Windward and Leeward Islands was much more general—the delivery of lectures on agricultural science to groups of schoolmasters to help them to take up nature study and to make the fullest use of school gardens.

Looking back I can now see where the emphasis of my job rightly lay. In Barbados I was a laboratory hermit, a specialist of specialists, intent on learning more and more about less and less: but in my tours of the various islands I was forced to forget my specialist studies and become interested in the growing of crops, which in these districts were principally cacao, arrowroot, ground nuts, sugar-cane, bananas, limes, oranges, and nutmegs. This contact with the land itself and with the practical men working on it laid the foundations of my knowledge of tropical agriculture.

This dual experience had not long been mine before I became aware of one disconcerting circumstance. I began to detect a fundamental weakness in the organization of that research which constituted officially the more important part of my work. I was an investigator of plant diseases, but I had myself no crops on which I could try out the remedies I advocated: I could not take my own advice before offering it to other people. It was borne in on me that there was a wide chasm between science in the laboratory and practice in the field, and I began to suspect that unless this gap could be bridged no real progress could be made in

1

the control of plant diseases: research and practice would remain apart: mycological work threatened to degenerate into little more than a convenient agency by which—provided I issued a sufficient supply of learned reports fortified by a judicious mixture of scientific jargon—practical difficulties could be side-tracked.

Towards the end of 1902, therefore, I took steps which terminated my appointment and gave me a fresh start. My next post was more promising—that of Botanist to the South-Eastern Agricultural College at Wye in Kent, where in addition to teaching I was placed in charge of the experiments on the growing and drying of hops which had been started by the former Principal, Mr. A. D. (later Sir Daniel) Hall. These experiments brought me in contact with a number of the leading hop growers, notably Mr. Walter (afterwards Sir Walter) Berry, Mr. Alfred Amos, and Colonel Honyball—all of whom spared no pains in helping me to understand the cultivation of this most interesting crop. I began to raise new varieties of hops by hybridization and at once made a significant practical discovery—the almost magical effect of pollination in speeding up the growth and also in increasing the resistance of the developing female flowers (the hops of commerce) to green-fly and mildew (a fungous disease) which often did considerable damage. The significant thing about this work was that I was meeting the practical men on their own ground. Actually their practice—that of eliminating the male plant altogether from their hop gardens—was a wide departure from natural law. My suggestion amounted to a demand that Nature be no longer defied. It was for this reason highly successful. By restoring pollination the health, the rate of growth, and finally the yield of hops were improved. Soon the growers all over the hop-growing areas of England saw to it that their gardens were provided with male hops, which liberated ample pollen just as it was needed.

This, my first piece of really successful work, was done during the summer of 1904—five years after I began research. It was obtained by happy chance and gave me a glimpse of the way Nature regulates her kingdom: it also did much to strengthen my conviction that the most promising method of dealing with plant diseases lay in prevention—by tuning up agricultural practice. But to continue such work the investigator would need land and hops of his own with complete freedom to grow them in his own way. Such facilities were not available and did not seem possible at Wye.

Then my chance came. Early in 1905 I was offered and accepted the post of Economic Botanist at the Agricultural Research Institute about to be founded by Lord Curzon, the then Viceroy of India, at Pusa in Bengal. On arrival in India in May 1905 the new institute only existed on paper, but an area of about seventy-five acres of land at one end of the Pusa Estate had not yet been allocated. I secured it instantly and spent my first five years in India learning how to grow the crops which it was my duty to improve by modern plant-breeding methods.

It was a decided advantage that officially my work was now no longer concerned merely with the narrow problem of disease. My main duties at Pusa were the improvement of crops and the production of new varieties. Over a period of nineteen years (1905–24) my time was devoted to this task, in the course of which many new types of wheat (including rust-resistant varieties), of tobacco, gram, and linseed were isolated, tested, and widely distributed.

In pursuance of the principle I had adopted of joining practice to my theory, the first step was to grow the crops I had to improve. I determined to do so in close conformity with local methods. Indian agriculture can point to a history of many centuries: there are records of the same rice fields being farmed in north-east India which go back for hundreds of years. What could be more sensible than to watch and learn from an experience which had passed so prolonged a test of time? I therefore set myself to make a preliminary study of Indian agriculture and speedily found my reward.

Now the crops grown by the cultivators in the neighbourhood of Pusa were remarkably free from pests: such things as insecticides and fungicides found no place in this ancient system of cultivation. This was a very striking fact, and I decided to break new ground and try out an idea which had first occurred to me in the West Indies and had forced itself on my attention at Wye, namely, to observe what happened when insect and fungous diseases were left alone and allowed to develop unchecked, indirect methods only, such as improved cultivation and more efficient varieties, being employed to prevent attacks.

In pursuit of this idea I found I could do no better than watch the operations of the peasants as aforesaid and regard them and the pests for the time being as my best instructors.

In order to give my crops every chance of being attacked by parasites nothing was done in the way of direct prevention; no insecticides and fungicides were used; no diseased material was ever destroyed. As my understanding of Indian agriculture progressed and as my practice im-

3

proved, a marked diminution of disease in my crops occurred. At the end of five years' tuition under my new professors—the peasants and the pests—the attacks of insects and fungi on all crops whose root systems suited the local soil conditions became negligible. By 1919 I had learnt how to grow healthy crops, practically free from disease, without the slightest help from mycologists, entomologists, bacteriologists, agricultural chemists, statisticians, clearing-houses of information, artificial manures, spraying machines, insecticides, fungicides, germicides, and all the other expensive paraphernalia of the modern experiment station.

This preliminary exploration of the ground suggested that the birth-right of every crop is health.

In the course of the cultivation of the seventy-five acres at my disposal I had to make use of the ordinary power unit in Indian agriculture which is oxen. It occurred to me that the same practices which had been so successful in the growing of my crops might be worth while if applied to my animals. To carry out such an idea it was necessary to have these work cattle under my own charge, to design their accommodation, and to arrange for their feeding, hygiene, and management. At first this was refused, but after persistent importunity backed by the powerful support of the Member of the Viceroy's Council in charge of Agriculture (the late Sir Robert Carlyle, K.C.S.I.), I was allowed to have charge of six pairs of oxen. I had little to learn in this matter, as I belong to an old agricultural family and was brought up on a farm which had made for itself a local reputation in the management of cattle. My work animals were most carefully selected and everything was done to provide them with suitable housing and with fresh green fodder, silage, and grain, all produced from fertile land. I was naturally intensely interested in watching the reaction of these well-chosen and well-fed oxen to diseases like rinderpest, septicaemia, and foot-and-mouth disease which frequently devastated the countryside.[1] None of my animals were segregated; none were inoculated; they frequently came in contact with diseased stock. As my small farmyard at Pusa was only separated by a low hedge from one of the large cattle-sheds on the Pusa estate, in which outbreaks of foot-and-mouth disease often occurred, I have several times seen my oxen rubbing noses with foot-and-mouth cases. Nothing happened. The healthy, well-fed animals failed to react to this disease exactly as suitable varieties of crops, when properly grown, did to insect

[1] These epidemics are the result of starvation, due to the intense pressure of the bovine population on the limited food supply.

4

and fungous pests—no infection took place. These experiences were afterwards repeated at Indore in Central India, but here I had forty, not twelve, oxen. A more detailed account of the prevention and cure of foot-and-mouth disease is given in a later chapter (p. 158).

These observations, important as they appeared both at the time and in retrospect, were however only incidental to my main work which was, as already stated, the improvement of the varieties of Indian crops, especially wheat. It was in the testing of the new kinds, which in the case of wheat soon began to spread over some millions of acres of India, that there gradually emerged the principle of which my observations about disease did but supply the first links in evidence: namely, that the foundations of all good cultivation lie not so much in the plant as in the soil itself: there is so intimate a connection between the state of soil, i.e. its fertility, and the growth and health of the plant as to outweigh every other factor. Thus on the capital point of increase of yield, if by improvement in selection and breeding my new special varieties of wheat, etc., might be estimated to produce an increase of 10 to 15 per cent, such yields could at once be increased not by this paltry margin, but doubled or even trebled, when the new variety was grown in soil brought up to the highest state of fertility. My results were afterwards amply confirmed by my colleague, the late Mr. George Clarke, C.I.E., who, by building up the humus content of his experiment station at Shahjahanpur in the United Provinces and by adopting simple improvements in cultivation and green-manuring, was able to treble the yields of sugarcane and wheat.

Between the years 1911 and 1918 my experience was considerably enlarged by the study of the problems underlying irrigation and fruit growing. For this purpose I was provided with a small experimental farm on the loess soils of the Quetta valley in Baluchistan where, till 1918, the summer months were spent. After a supply of moisture had been provided to supplement the scanty winter rainfall, the limiting factors in crop production proved to be soil aeration and the humus content of the land. Failure to maintain aeration was indicated by a disease of the soil itself. The soil flora became anaerobic: alkali salts developed: the land died. The tribesmen kept the alkali condition at bay in their fruit orchards in a very suggestive manner—by means of the deep-rooting system of lucerne combined with surface dressings of farmyard manure. Moreover they invariably combined their fruit growing with mixed farming and livestock. Nowhere, as in the West, did one

5

find the whole farm devoted to fruit with no provision for an adequate supply of animal manure. This method of fruit growing was accompanied by an absence of insect and fungoid diseases: spraying machines and poison sprays were unheard of: artificial manures were never used. The local methods of grape growing were also intensely interesting. To save the precious irrigation water and as a protection from the hot, dry winds, the vines were planted in narrow ditches dug on the slopes of the valley and were always manured with farmyard manure. Irrigation water was led along the ditches and the vines were supported by the steep sides of the trenches. At first sight all the conditions for insect and fungous diseases seemed to be provided, but the plants were remarkably healthy. I never found even a trace of disease. The quality of the produce was excellent: the varieties grown were those which had been in cultivation in Afghanistan for centuries. No signs of running out were observed. Here were results in disease resistance and in the stability of the variety in striking contrast to those of western Europe, where disease is notorious, the use of artificial manures and poison sprays is universal, and where the running out of the variety is constantly taking place (see also p. 132).

These results and observations taken together and prolonged over a period of nineteen years at length indicated what should be the right method of approach to the work I was doing. Improvement of varieties, increased yields, freedom from disease were not distinct problems, but formed parts of one subject and, so to speak, were members one of another, all arising out of the great linkage between the soil, the plant, and the animal. The line of advance lay not in dealing with these factors separately but together. If this were to be the path of progress and if it was useless to proceed except on the basis of crops grown on fertile land, then the first prerequisite for all subsequent work would be just the bringing of the experiment station area to the highest state of fertility and maintaining it in that condition.

This, however, opened up a further problem. The only manure at the command of the Indian cultivator was farmyard manure. Farmyard manure was therefore essential, but even on the experiment stations the supply of this material was always insufficient. The problem was how to increase it in a country where a good deal of the cattle-dung has to be burnt for fuel. No lasting good could be achieved unless this problem were overcome, for no results could be applied to the country at large.

The solution was suggested by the age-long practices of China, where a system of utilizing farm wastes and turning them into humus had been evolved which, if applied to India, would make every Indian holding self-supporting as regards manure. This idea called for investigation.

I now came up against a very great difficulty. Such a problem did not fall within my official sphere of work. It obviously necessitated a great deal of chemical and agricultural investigation under my personal control and complete freedom to study all aspects of the question. But while my idea was taking shape, the organization of agricultural research at Pusa had also developed. A series of watertight compartments—plant breeding, mycology, entomology, bacteriology, agricultural chemistry, and practical agriculture—had become firmly established. Vested interests were created which regarded the organization as more important than its purpose. There was no room in it for a comprehensive study of soil fertility and its many implications by one member of the staff with complete freedom of action. My proposals involved "overlapping," a defect which was anathema both to the official mind (which controlled finance) and to a research institute subdivided as Pusa always had been.

The obvious course was to leave the institute and to collect the funds to found a new centre where I could follow the gleam unhampered and undisturbed. After a delay of six precious years, 1918–24, the Indore Institute of Plant Industry (at which cotton was the principal crop) was founded, where I was provided with land, ample money, and complete freedom. Now the fundamental factor underlying the problems of Indian cotton was none other than the raising of soil fertility. I might therefore kill two birds with one stone. I could solve the cotton problem if I could increase the amount of farmyard manure for India as a whole.

At Indore I had a considerably larger area at my disposal, namely, 300 acres. From the outset the principles which I had worked out at Pusa were applied to 'cotton. The results were even better. The yield of cotton was almost trebled and the whole experiment station area stood out from the surrounding countryside by reason of the fine crops grown. Moreover these crops were free from disease, with only two exceptions, during the whole eight years of my work there, exceptions in themselves highly significant. A small field of gram, which had become accidentally water-logged three months before the crop was sown, was, a month after sowing, found to be heavily attacked by the gram caterpillar, the infected areas corresponding with the waterlogged areas with great exactness, while the rest of the plot remained unaffected:

the caterpillar did not spread, though nothing was done to check it. In the second case a field of *san* hemp (*Crotalaria juncea*, L.), originally intended for greenmanuring, was allowed to flower for seed; after flowering it was smothered in mildew and insect pests and no seed set. Subsequent trials showed that this crop will set seed and be disease free on black soils only if the land is previously well manured with farmyard manure or compost.

These results were progressive confirmation of the principle I was working out—the connection between land in good heart and disease-free crops: they were proof that as soon as land drops below par, disease may set in. The first case showed the supreme importance of keeping the physical texture of the soil right, the second was an interesting example of the refusal of Mother Earth to be overworked, of her unbreakable rule to limit herself strictly to that volume of operations for which she has sufficient reserves: flowers were formed, but seed refused to set and the mildew and insects were called in to remove the imperfect product.

These were the exceptions to prove the rule, for during the eight years of my work at Indore it was assumed by me as a preliminary condition to all experiments that my fields must be fertile. This was brought about by supplying them with heavy dressings of compost made on a simple development of the Chinese system. As I was now free, it was possible for me to make these arrangements on a large scale, and in the course of doing so it seemed well worth while to work out the theory that underlay the empiric Chinese practice. A complete series of experiments and investigations were carried out, establishing the main chemical, physical, and biological processes which go to humus formation in the making of compost. In this work I received valuable help from Mr. Y. D. Wad who was in charge of the chemical side cf the investigation. On my retirement from official service in 1931 I assumed that the publication of this joint work in book form would be the last scientific task which I should ever undertake.

It proved instead to be the beginning of a new period which has been based on the long preparation which preceded it: the years of work and experiment carried out in the tropics had gradually but inevitably led me up to the threshold of ideas which embrace and explain the facts and the practices, the theory and also the failures, which had met me in the course of these thirty-two years. Our book on *The Waste Products of Agriculture: Their Utilization as Humus,* designed to be a practical guide to assist the Indian cotton cultivators, evoked a much wider interest. The so-called Indore Process of making compost was started at a number of centres in other countries and interesting results began to be reported, very much like what I had obtained at Indore.

Two years after publication, in February 1933, I saw the inception of a compost-making scheme at Colonel Grogan's estate not far from Nairobi in Kenya Colony. During this visit it first occurred to me gradually to terminate all my other activities and to confine myself to encouraging the pioneers engaged in agriculture all over the world to restore and maintain the fertility of their land. This would involve a campaign to be carried out single-handed at my own expense as no official funds could be expected for a project such as mine. Even if I could have obtained the means needed it would have been necessary to work with research organizations I had long regarded not only as obsolete, but as the perfect means of preventing progress. A soil fertility campaign carried on by a retired official would also throw light on another question, namely, the relative value of complete freedom and independence in getting things done in farming, as compared with the present cumbrous and expensive governmental organization.

By the end of 1933 matters had progressed far enough to introduce the Indore Process to a wider public. This was done by means of two lectures before the Royal Society of Arts in 1933 and 1935, some thousands of extra copies of both of which were distributed all over the world, and subsequent contributions to the *Journal* of that society, to a German periodical—*Der Tropenpflanzer*—and a Spanish review—the *Revista del Instituto de Defensa del Café* of Costa Rica. The process became generally known and was found to be a most advantageous proposition in the big plantation industries—coffee, tea, sugar, maize, tobacco, sisal, rice, and vine—yields and quality alike being notably improved. I devoted my energies to advising and assisting those interested, and during this period became greatly indebted to the tea industry for material help and encouragement.

In 1937 results were reported in the case of tea which were difficult to explain. Single light dressings of Indore compost improved the yield of leaf and increased the resistance of the bush to insect attacks in a way which much surpassed what was normally to be expected from a first application. While considering these cases I happened to read an account of Dr. Rayner's work on conifers at Wareham in Dorsetshire, where small applications of humus had also produced spectacular results. Normally humus is considered to act on the plant indirectly: the

9

oxidation of the substances composing it ultimately forming salts in the soil, which are then absorbed by the root hairs in the usual processes of nutrition. Was there here, however, something more than this, some direct action having an immediate effect and one very powerful?

Such indeed has proved to be the case and the explanation can now be set forth of the wonderful double process by which Nature causes the plant to draw its nurture from the soil. The mechanism by which living fungous threads (mycelium) invade the cells of the young roots and are gradually digested by these is described in detail in a later chapter (p. 23). It was this, the mycorrhizal association, which was the explanation of what had happened to the conifers and the tea shrubs, both forest plants, a form of vegetation in which this association of root and fungus has been known for a long time. This direct method of feeding would account for the results observed.

A number of inquiries which I was now able to set on foot revealed the existence of this natural feeding mechanism in plant after plant, where it had hitherto neither been observed nor looked for, but only, be it noted, where there was ample humus in the soil. Where humus was wanting, the mechanism was either absent or ineffective: the plant was limited to the nurture derived by absorption of the salts in the soil solution: it could not draw on these rich living threads, abounding in protein.

The importance of the opening up of this aspect of plant nutrition was quite obvious. Here at last was a full and sufficient explanation of the facts governing the health of plants. From this point on evidence began to accumulate to illumine the new path of inquiry, which in my opinion is destined to lead us a very long way indeed. It was clear that the doubling of the processes of plant nutrition was one of those reserve devices on which rests the permanence and stability of Nature. Plants deprived of the mycorrhizal association continue to exist, but they lose both their power to resist shock and their capacity to reproduce themselves. A new set of facts suddenly fell into place: the running out of varieties, a marked phenomenon of modern agriculture, to answer which new varieties of the important crops have constantly to be bred— hence the modern plant breeding station—could without hesitation be attributed to the continued impoverishment of modern soils owing to the prolonged negligence of the Western farmer to feed his fields with humus. By contrast the maintenance of century-old varieties in the East, so old that in India they bear ancient Sanskrit names, was proof of the unimpaired capacity of the plant to breed in those countries where humus was abundantly supplied.

The mycorrhizal association may not prove to be the only path by which the nitrogen complexes derived from the digestion of proteins reach the sap. Humus also nourishes countless millions of bacteria whose dead bodies leave specks of protein thickly strewn throughout the soil. But these complex bodies are not permanent: they are reduced by other soil organisms to simpler and simpler bodies which finally become mineralized to form the salts taken up by the roots for use in the green leaves. May not some of the very early stages in the oxidation of these specks of protein be absorbed by the root hairs from the soil water? It would seem so, because a few crops exist, like the tomato, which although reacting to humus are not provided with the mycorrhizal association. This matter is discussed in the next chapter (p. 17).

These results set up a whole train of thought. The problem of disease and health took on a wider scope. In March 1939, new ground was broken. The Local Medical and Panel Committees of Cheshire, summing up their experience of the working of the National Health Insurance Act for over a quarter of a century in the county, did not hesitate to link up their judgment on the unsatisfactory state of health of the human population under their care with the problem of nutrition, tracing the line of fault right back to an impoverished soil and supporting their contentions by reference to the ideas which I had for some time been advocating. Their arguments were powerfully supported by the results obtained at the Peckham Health Centre and by the work, already published, of Sir Robert McCarrison, which latter told the story from the other side of the world and from a precisely opposite angle—he was able to instance an Eastern people, the Hunzas, who were the direct embodiment of an ideal of health and whose food was derived from soil kept in a state of the highest natural fertility.

By these contemporaneous pioneering efforts the way was blazed for treating the whole problem of health in soil, plant, animal, and man as one great subject, calling for a boldly revised point of view and entirely fresh investigations.

By this time sufficient evidence had accumulated for setting out in book form the case for soil fertility. This was published in June 1940 by the Oxford University Press under the title of *An Agricultural Testament*. This book, now in its fourth English and second American edition, set forth the whole gamut of connected problems as far as can at

11

present be done—what wider revelations the future holds is not yet fully disclosed. In it I summed up my life's work and advanced the following views:

1. The birthright of all living things is health.

2. This law is true for soil, plant, animal, and man: the health of these four is one connected chain.

3. Any weakness or defect in the health of any earlier link in the chain is carried on to the next and succeeding links, until it reaches the last, namely, man.

4. The widespread vegetable and animal pests and diseases, which are such a bane to modern agriculture, are evidence of a great failure of health in the second (plant) and third (animal) links of the chain.

5. The impaired health of human populations (the fourth link) in modern civilized countries is a consequence of this failure in the second and third links.

6. This general failure in the last three links is to be attributed to failure in the first link, the soil: the undernourishment of the soil is at the root of all. The failure to maintain a healthy agriculture has largely cancelled out all the advantages we have gained from our improvements in hygiene, in housing, and our medical discoveries.

7. To retrace our steps is not really difficult if once we set our minds to the problem. We have to bear in mind Nature's dictates, and we must conform to her imperious demand: (a) for the return of all wastes to the land; (b) for the mixture of the animal and vegetable existence; (c) for the maintaining of an adequate reserve system of feeding the plant, i.e. we must not interrupt the mycorrhizal association. If we are willing so far to conform to natural law, we shall rapidly reap our reward not only in a flourishing agriculture, but in the immense asset of an abounding health in ourselves and in our children's children.

These ideas, straightforward as they appear when set forth in the form given above, conflict with a number of vested interests. It has been my self-appointed task during the last few years of my life to join hands with those who are convinced of their truth to fight the forces impeding progress. So large has been the flow of evidence accumulating that in 1941 it was decided to publish a *News-Letter on Compost,* embodying the most interesting of the facts and opinions reaching me or others in the campaign. The *News-Letter,* which appears three times a year under the aegis of the Cheshire Local Medical and Panel Committees, has grown from eight to sixty-four pages and is daily gaining new readers.

The general thesis that no one generation has a right to exhaust the soil from which humanity must draw its sustenance has received further powerful support from religious bodies. The clearest short exposition of this idea is contained in one of the five fundamental principles adopted by the recent Malvern Conference of the Christian Churches held with the support of the late Archbishop of Canterbury, Dr. Temple. It is as follows: "The resources of the earth should be used as God's gifts to the whole human race and used with due consideration for the needs of the present and future generations."

Food is the chief necessity of life. The plans for social security which are now being discussed merely guarantee to the population a share in a variable and, in present circumstances, an uncertain quantity of food, most of it of very doubtful quality. Real security against want and ill health can only be assured by an abundant supply of fresh food properly grown in soil in good heart. *The first place in post-war plans of reconstruction must be given to soil fertility in every part of the world.* The land of this country and the Colonial Empire, which is the direct responsibility of Parliament, must be raised to a higher level of productivity by a rational system of farming which puts a stop to the exploitation of land for the purpose of profit and takes into account the importance of humus in producing food of good quality. The electorate alone has the power of enforcing this and to do so it must first realize the full implications of the problem.

They and they alone possess the power to insist that every boy and every girl shall enter into their birthright—health, and that efficiency, well-being, and contentment which depend thereon. One of the objects of this book is to show the man in the street how this world of ours can be born again. He can help in this task, which depends at least as much on the plain efforts of the plain man in his own farm, garden, or allotment as on all the expensive paraphernalia, apparatus, and elaboration of the modern scientist: more so in all probability, inasmuch as one small example always outweighs a ton of theory. If this sort of effort can be made and the main outline of the problems at stake are grasped, nothing can stop an immense advance in the well-being of our earth. A healthy population will be no mean achievement, for our greatest possession is ourselves.

The man in the street will have to do three things:

1. He must create in his own farm, garden, or allotment examples without end of what a fertile soil can do.

13

2. He must insist that the public meals in which he is directly interested, such as those served in boarding schools, in the canteens of day schools and of factories, in popular restaurants and tea shops, and at the seaside resorts at which he takes his holidays are composed of the fresh produce of fertile soil.

3. He must use his vote to compel his various representatives—municipal, county, and parliamentary—to see to it: (*a*) that the soil of his country is made fertile and maintained in this condition: (*b*) that the public health system of the future is based on the fresh produce of land in good heart.

This introduction started with the training of an agricultural investigator: it ends with the principles underlying the public health system of to-morrow. It has, therefore, covered much ground in describing what is nothing less than an adventure in scientific research. One lesson must be stressed: The difficulties met with and overcome in the official portion of this journey were not part of the subject investigated. They were man made and created by the research organization itself. More time and energy had to be expended in side-tracking the lets and hindrances freely strewn along the road by the various well-meaning agencies which controlled discovery than in conducting the investigations themselves. When the day of retirement came, all these obstacles vanished and the delights of complete freedom were enjoyed. Progress was instantly accelerated. Results were soon obtained throughout the length and breadth of the English-speaking world which make crystal clear the great role which soil fertility must play in the future of mankind.

The real Arsenal of Democracy is a fertile soil, the fresh produce of which is the birthright of the nations.

PART I

THE PART PLAYED BY SOIL FERTILITY IN AGRICULTURE

2

THE OPERATIONS OF NATURE

THE INTRODUCTION to this book describes an adventure in agricultural research and records the conclusions reached. If the somewhat unorthodox views set out are sound, they will not stand alone but will be supported and confirmed in a number of directions—by the farming experience of the past and above all by the way Nature, the supreme farmer, manages her kingdom. In this chapter the manner in which she conducts her various agricultural operations will be briefly reviewed. In surveying the significant characteristics of the life—vegetable and animal—met with in Nature particular attention will be paid to the importance of fertility in the soil and to the occurrence and elimination of disease in plants and animals.

What is the character of life on this planet? What are its great qualities? The answer is simple: The outstanding characteristics of Nature are variety and stability.

The variety of the natural life around us is such as to strike even the child's imagination, who sees in the fields and copses near his home, in the ponds and streams and seaside pools round which he plays, or, if being city-born he be deprived of these delightful playgrounds, even in his poor back-garden or in the neighbouring park, an infinite choice of different flowers and plants and trees, coupled with an animal world full of rich changes and surprises, in fact, a plenitude of the forms of living things constituting the first and probably the most powerful introduction he will ever receive into the nature of the universe of which he is himself a part.

The infinite variety of forms visible to the naked eye is carried much farther by the microscope. When, for example, the green slime in stagnant water is examined, a new world is disclosed—a multitude of simple flowerless plants—the blue-green and the green algae—always accompanied by the lower forms of animal life. We shall see in a later chapter (p. 127) that on the operations of these green algae the well-being of the rice crop, which nourishes countless millions of the human race, depends. If a fragment of mouldy bread is suitably magnified, members of

17

still another group of flowerless plants, made up of fine, transparent threads entirely devoid of green colouring matter, come into view. These belong to the fungi, a large section of the vegetable kingdom, which are of supreme importance in farming and gardening.

It needs a more refined perception to recognize throughout this stupendous wealth of varying shapes and forms the principle of stability. Yet this principle dominates. It dominates by means of an ever-recurring cycle, a cycle which, repeating itself silently and ceaselessly, ensures the continuation of living matter. This cycle is constituted of the successive and repeated processes of birth, growth, maturity, death, and decay.

An eastern religion calls this cycle the Wheel of Life and no better name could be given to it. The revolutions of this Wheel never falter and are perfect. Death supersedes life and life rises again from what is dead and decayed.

Because we are ourselves alive we are much more conscious of the processes of growth than we are of the processes involved in death and decay. This is perfectly natural and justifiable. Indeed, it is a very powerful instinct in us and a healthy one. Yet, if we are fully grown human beings, our education should have developed in our minds so much of knowledge and reflection as to enable us to grasp intelligently the vast role played in the universe by the processes making up the other or more hidden half of the Wheel. In this respect, however, our general education in the past has been gravely defective partly bcause science itself has so sadly misled us. Those branches of knowledge dealing with the vegetable and animal kingdoms—botany and zoology— have confined themselves almost entirely to a study of *living* things and have given little or no attention to what happens to these units of the universe when they die and to the way in which their waste products and remains affect the general environment on which both the plant and animal world depend. When science itself is unbalanced, how can we blame education for omitting in her teaching one of the things that really matter?

For though the phases which are preparatory to life are, as a rule, less obvious than the phases associated with the moment of birth and the periods of growth, they are not less important. If once we can grasp this and think in terms of ever-repeated advance and recession, recession and advance, we have a truer view of the universe than if we define death merely as an ending of what has been alive.

18

Nature herself is never satisfied except by an even balancing of her processes—growth and decay. It is precisely this even balancing which gives her unchallengeable stability. That stability is rock-like. Indeed, this figure of speech is a poor one, for the stability of Nature is far more permanent than anything we can call a rock—rocks being creations which themselves are subject to the great stream of dissolution and rebirth, seeing that they suffer from weathering and are formed again, that they can be changed into other substances and caught up in the grand process of living: they too, as we shall see (p. 85), are part of the Wheel of Life. However, we may at a first glance omit the changes which affect the inert masses of this planet, petrological and mineralogical: though very soon we shall realize how intimate is the connection even between these and what is, in the common parlance, alive. There is a direct bridge between things inorganic and things organic and this too is part of the Wheel.

But before we start on our examination of that part of the great process which now concerns us—namely, plant and animal life and the use man makes of them—there is one further idea which we must master. It is this: The stability of Nature is secured not only by means of a very even balancing of her Wheel, by a perfect timing, so to say, of her mechanisms, but also rests on a basis of enormous reserves. Nature is never a hand-to-mouth practitioner. She is often called lavish and wasteful, and at first sight one can be bewildered and astonished at the apparent waste and extravagance which accompany the carrying on of vegetable and animal existence. Yet a more exact examination shows her working with an assured background of accumulated reserves, which are stupendous and also essential. The least depletion in these reserves induces vast changes and not until she has built them up again does she resume the particular process on which she was engaged. A realization of this principle of reserves is thus a further necessary item in a wide view of natural law. Anyone who has recovered from a serious illness, during which the human body lives partly on its own reserves, will realize how Nature afterwards deals with such situations. During the period of convalescence the patient appears to make little progress till suddenly he resumes his old-time activities. During this waiting period the reserves used up during illness are being replenished.

THE LIFE OF THE PLANT

A survey of the Wheel of Nature will best start from that rather rapid series of processes which cause what we commonly call living matter to come into active existence; that is, in fact, from the point where life most obviously, to our eyes, begins. The section of the Wheel embracing these processes is studied in physiology from the Greek word φύσις, the root φύω meaning to bring to life, to grow.

But how does life begin on this planet? We can only say this: that the prime agency in carrying it on is sunlight, because it is the source of energy, and that the instrument for intercepting this energy and turning it to account is the green leaf.

This wonderful little example of Nature's invention is a battery of intricate mechanisms. Each cell in the interior of a green leaf contains minute specks of a substance called chlorophyll and it is this chlorophyll which enables the plant to grow. Growth implies a continuous supply of nourishment. Now plants do not merely collect their food: they manufacture it before they can feed. In this they differ from animals and man, who search for what they can pass through their stomachs and alimentary systems, but cannot do more; if they are unable to find what is suitable to their natures and ready for them, they perish. A plant is, in a way, a more wonderful instrument. It is an actual food factory, making what it requires before it begins the processes of feeding and digestion. The chlorophyll in the green leaf, with its capacity for intercepting the energy of the sun, is the power unit that, so to say, runs the machine. The green leaf enables the plant to draw simple raw materials from diverse sources and to work them up into complex combinations.

Thus from the air it absorbs carbon-dioxide (a compound of two parts of oxygen to one of carbon), which is combined with more oxygen from the atmosphere and with other substances, both living and inert, drawn from the soil and from the water which permeates the soil. All these raw materials are then assimilated in the plant and made into food. They become organic compounds, i.e. compounds of carbon, classified conveniently into groups known as carbohydrates, proteins, and fats; together with an enormous volume of water (often over 90 per cent of the whole plant) and interspersed with small quantities of chemical salts which have not yet been converted into the organic phase, they make up the whole structure of the plant—root, stem, leaf, flower, and seed. This structure includes a big food reserve. The life principle,

PLATE I. OBSERVATION CHAMBER FOR ROOT STUDIES
AT EAST MALLING

PLATE II. THE BEGINNINGS OF MYCORRHIZAL
ASSOCIATION IN THE APPLE

Root-tip (× 12) of *Lane's Prince Albert* on root-stock M *XVI* at sixteen inches below the surface, showing root-cap (A), young root hairs (C), and older root hairs with drops of exudate (C_1). The cobweb-like mycelial strands are well seen approaching the rootlet in the region marked (C).

the nature of which evades us and in all probability always will, resides in the proteins looked at in the mass. These proteins carry on their work in a cellulose framework made up of cells protected by an outer integument and supported by a set of structures known as the vascular bundles, which also conduct the sap from the roots to the leaves and distribute the food manufactured there to the various centres of growth. The whole of the plant structures are kept turgid by means of water.

The green leaf, with its chlorophyll battery, is therefore a perfectly adapted agency for continuing life. It is, speaking plainly, the only agency that can do this and is unique. Its efficiency is of supreme importance. Because animals, including man, feed eventually on green vegetation, either directly or through the bodies of other animals, it is our sole final source of nutriment. There is no alternative supply. Without sunlight and the capacity of the earth's green carpet to intercept its energy for us, our industries, our trade, and our possessions would soon be useless. It follows therefore that everything on this planet must depend on the way mankind makes use of this green carpet, in other words on its efficiency.

The green leaf does not, however, work by itself. It is only a part of the plant. It is curious how easy it is to forget that normally we see only one-half of each flowering plant, shrub, or tree: the rest is buried in the ground. Yet the dying down of the visible growth of many plants in the winter, their quick reappearance in the spring, should teach us how essential and important a portion of all vegetation lives out of our sight; it is evident that the root system, buried in the ground, also holds the life of the plant in its grasp. It is therefore not surprising to find that leaves and roots work together, forming a partnership which must be put into fresh working order each season if the plant is to live and grow.

If the function of the green leaf armed with its chlorophyll is to manufacture the food the plant needs, the purpose of the roots is to obtain the water and most of the raw materials required—the sap of the plant being the medium by which these raw materials (collected from the soil by the roots) are moved to the leaf. The work of the leaf we found to be intricate: that of the roots is not less so. What is surprising is to come upon two quite distinct ways in which the roots set about collecting the materials which it is their business to supply to the leaf; these two methods are carried on simultaneously. We can make a very shrewd guess at the master principle which has put the second method alongside the first: it is again the principle of providing a reserve—this time of the vital proteins.

None of the materials that reach the green leaf by whatever method is food: it is only the raw stuff from which food can be manufactured. By the first method, which is the most obvious one, the root hairs search out and pass into the transpiration current of the plant dissolved substances which they find in the thin films of water spread between and around each particle of earth; this film is known as the soil solution. The substances dissolved in it include gases (mainly carbon dioxide and oxygen) and a series of other substances known as chemical salts like nitrates, compounds of potassium and phosphorus, and so forth, all obtained by the breaking down of organic matter or from the destruction of the mineral portions of the soil. In this breaking down of organic matter we see in operation the reverse of the building-up process which takes place in the leaf. Organic matter is continuously reverting to the inorganic state: it becomes mineralized: nitrates are one form of the outcome. It is the business of the root hairs to absorb these substances from the soil solution and to pass them into the sap, so that the new life-building process can start up again. In a soil in good heart the soil solution will be well supplied with these salts. Incidentally we may note that it has been the proved existence of these mineral chemical constituents in the soil which, since the time of Liebig, has focused attention on soil chemistry and has emphasized the passage of chemical food materials from soil to plant to the neglect of other considerations.

But the earth's green carpet is not confined to its remarkable power of transforming the inert nitrates and mineral contents of the soil into an active organic phase: it is utilized by Nature to establish for itself, in addition, a direct connection, a kind of living bridge, between its own life and the living portion of the soil. This is the second method by which plants feed themselves. The importance of this process, physiological in nature and not merely chemical, cannot be over-emphasized and some description of it will now be attempted.

THE LIVING SOIL

The soil is, as a matter of fact, full of live organisms. It is essential to conceive of it as something pulsating with life, not as a dead or inert mass. There could be no greater misconception than to regard the earth

22

as dead: a handful of soil is teeming with life. The living fungi, bacteria, and protozoa, invisibly present in the soil complex, are known as the soil population. This population of millions and millions of minute existences, quite invisible to our eyes of course, pursue their own lives. They come into being, grow, work, and die: they sometimes fight each other, win victories, or perish; for they are divided into groups and families fitted to exist under all sorts of conditions. The state of a soil will change with the victories won or the losses sustained; and in one or other soil, or at one or other moment, different groups will predominate.

This lively and exciting life of the soil is the first thing that sets in motion the great Wheel of Life. Not without truth have poets and priests paid worship to "Mother Earth," the source of our being. What poetry or religion have vaguely celebrated, science has minutely examined, and very complete descriptions now exist of the character and nature of the soil population, the various species of which have been classified, labelled, and carefully observed. It is this life which is continually being passed into the plant.

The process can actually be followed under the microscope. Some of the individuals belonging to one of the most important groups in this mixed population—the soil fungi—can be seen functioning. If we arrange a vertical darkened glass window on the side of a deep pit in an orchard, it is not difficult to see with the help of a good lens or a low-power horizontal microscope (arranged to travel up and down a vertical fixed rod) some of these soil fungi at work. They are visible in the interstices of the soil as glistening white branching threads, reminiscent of cobwebs. In Dr. Rogers's interesting experiments on the root systems of fruit trees at East Malling Research Station, where this method of observing them was initiated and demonstrated to me, these fungous threads could be seen approaching the young apple roots in the absorbing region (just behind the advancing root tips) on which the root hairs are to be found. Dr. Rogers very kindly presented me with two excellent photographs—one showing the general arrangement of his observation chamber (Plate I), the other, taken on 6th July 1933, of a root tip (magnified by about twelve) of *Lane's Prince Albert* (grafted on root stock XVI) at sixteen inches below the surface, showing abundant fungous strands running in the soil and coming into direct contact with the growing root (Plate II).

But this is only the beginning of the story. When a suitable section of one of these young apple roots, growing in fertile soil and bearing active

23

root hairs, is examined, it will be found that these fine fungous threads actually invade the cells of the root, where they can easily be observed passing from one cell to another. But they do not remain there very long. After a time the apple roots absorb these threads. All stages of the actual digestion can be seen.

The significance of this process needs no argument. Here we have a simple arrangement on the part of Nature by which the soil material on which these fungi feed can be joined up, as it were, with the sap of the tree. These fungous threads are very rich in protein and may contain as much as 10 per cent of organic nitrogen; this protein is easily digested by the ferments (enzymes) in the cells of the root; the resulting nitrogen complexes, which are readily soluble, are then passed into the sap current and so into the green leaf. An easy passage, as it were, has been provided for food material to move from soil to plant in the form of proteins and their digestion products, which latter in due course reach the green leaf. The marriage of a fertile soil and the tree it nourishes is thus arranged. Science calls these fungous threads *mycelium* (again from a Greek word, μύκης), and as the Greek for root is ῥίζα (*rhiza*, cf. rhizome), the whole process is known as the *mycorrhizal association*. This partnership is universal in the forest and is general throughout the vegetable kingdom. A few exceptions, however, exist which will be referred to in the next paragraph.

Among the plants in which this mycorrhizal association has hitherto not been observed are the tomato and certain cultivated members of the cabbage family, many of which possess a very diffuse root system and exceptionally elongated root hairs. Nevertheless, all these examples respond very markedly to the condition of the soil in which they are grown and if fed with dressings of humus will prosper. The question naturally arises: Exactly how does this take place? What is the alternative mechanism that replaces the absent mycorrhizal association?

A simple explanation would appear to be this. Fertile soils invariably contain a greatly enhanced bacterial population whose dead remains must be profusely scattered in the water films which bathe the com-

---

[1] The reader who wishes to delve into the technical details relating to the mycorrhizal association and its bearing on forestry and agriculture should consult the following works:—

1. Rayner, M. C. and Neilson-Jones, W.—*Problems in Tree Nutrition*, Faber & Faber, London, 1944.
2. Balfour, Lady Eve—*The Living Soil*, Faber & Faber, London, 1944.
3. Howard, Sir Albert—*An Agricultural Testament*, Oxford University Press, 1940.
4. Rodale, J. I.—*Pay Dirt*, The Devin-Adair Company, New York, 1945.

pound soil particles and the root hairs of the crops themselves; these specks of dead organic matter, rich in protein, are finally mineralized into simple salts like nitrates. We have already mentioned this breaking-down process of the soil population. What is here to be noted is that it is no sudden transformation, but takes place in stages. May not, therefore, some at least of the first-formed nitrogen complexes, which result from this breaking down, be absorbed by the root hairs and so added to the sap current? That is to say that the non-mycorrhiza-forming plants, not drawing on the soil fungi, do compensate themselves by absorbing organic nitrogen in this form—they catch the bacterial soil population, as it were, before it has been reduced to an entirely inert phase and so have their link also with the biological life of the soil. That there must be some such passage of matter on a biological basis is suggested by the fact that only in fertile soil, i.e. in soils teeming with bacteria, do these non-mycorrhiza formers reveal resistance to disease and high quality in the produce, which means that only in these soils are they really properly fed.

This would be a third method used by plants for feeding themselves, a sort of half-way method between the absorption powers exercised by the root hairs and the direct digestive capacity of the roots: as the mechanism used in this method is presumably the root hairs, the diffuseness of the root system of plants of the cabbage family would be explained. It is possible that even mycorrhiza formers use this alternative passage for organic nitrogen. There seems no reason at all why this should not be so.

But how do the various agencies concerned in these intricate operations manage to carry on their work, buried as they are away from the light and thus unable to derive anything from the source of energy, the sun? How do they do their initial work at all until they can hand over to the green leaf? They derive their energy by oxidising (i.e. burning up) the stores of organic matter in the soil. As in an ordinary fire, this process of oxidation releases energy. The oxygen needed for this slow combustion is drawn from the air, in part washed down by the rain, which dissolves it from the atmosphere in its descent. Incidentally this explains why rain is so superior as a moistening agency for plants to any form of watering from a can: incidentally, again, we can understand the need for cultivating the soil and keeping it open, so that the drawing in of oxygen, or the respiration of the soil, can proceed and the excess carbon dioxide can be expelled into the atmosphere.

Humus is the Latin word for soil or earth. But as used by the husbandman humus nowadays does not mean just earth in general, but indicates that undecayed residue of vegetable and animal waste lying on the surface, combined with the dead bodies of these bacteria and fungi themselves when they have done their work, the whole being a highly complex and somewhat varying substance which is, so to say, the mine or store or bank from which the organisms of the soil and then, in direct succession, the plant, the tree, and thereafter the animal, draw what they need for their existence. This store is all important.

THE SIGNIFICANCE OF HUMUS

Humus is the most significant of all Nature's reserves and as such deserves a detailed examination.

A very perfect example of the methods by which Nature makes humus and thus initiates the turning of her Wheel is afforded by the floor of the forest. Dig down idly with a stick under any forest tree: first there will be a rich, loose accumulation of litter made up of dead leaves, flowers, twigs, fragments of bark, bits of decaying wood, and so forth, passing gradually as the material becomes more tightly packed into rich, moist, sweet-smelling earth, which continues downwards for some inches and which, when disturbed, reveals many forms of tiny insect and animal life. We have been given here a glimpse of the way Nature makes humus—the source from which the trunk of the tree has drawn its resisting strength, its leaves their glittering beauty.

Throughout the year, endlessly and continuously, though faster at some seasons than at others, the wastes of the forest thus accumulate and at once undergo transformation. These wastes are of many kinds and mix as they fall; for leaf mingles with twig and stem, flower with moss, and bark with seed-coats. Moreover, vegetable mingles with animal. Let us beware of the false idea that the forest is a part of the vegetable kingdom only. Millions of animal existences are housed in it; mammals and birds are everywhere and can be seen with the naked eye. The lower forms of animal life—the invertebrates—are even more numerous. Insects, earthworms, and so forth are obvious: the microscope reveals new worlds of animal life down to simple protozoa. The excreta of these animals while living and their dead bodies constitute an important component of what lies on the forest floor; even the bodies of in-

26

sects form in the mass a constituent element not without importance, so that in the end the two sources of waste are completely represented and are, above all, completely mingled. But the volume of the vegetable wastes is several times greater than that of the animal residues.

These wastes lie gently, only disturbed by wind or by the foot of a passing animal. The top layer is thus very loose; ample air circulates for several inches downwards: the conditions for the fermentation by the moulds and microbes (which feed on the litter) are, as the scientist would say, *aerobic*. But partly by pressure from above and partly as the result of fermentation the lower layers are forced to pack more closely and the final manufacture of humus goes on without much air: the conditions are now *anaerobic*. This is a succession of two modes of manufacture which we shall do well to remember, as in our practical work it has to be imitated (p. 217).

This mass of accumulated wastes is acted on by the sunlight and the rain; both are dispersed and fragmented by the leaf canopy of the trees and undergrowth. The sunlight warms the litter; the rain keeps it moist. The rain does not reach the litter as a driving sheet, but is split up into small drops the impetus of whose fall is well broken. Nor does the sunlight burn without shade; it is tempered. Finally, though air circulates freely, there is perfect protection from the cooling and drying effects of strong wind.

With abundant air, warmth, and water at their disposal the fungi and bacteria, with which, as we have already noted, the soil is teeming, do their work. The fallen mixed wastes are broken up; some passes through the bodies of earthworms and insects: all is imperceptibly crumbled and changed until it decomposes into that rich mass of dark colour and earthy smell which is so characteristic of the forest floor and which holds such a wealth of potential plant nourishment.

The process that takes place in a prairie, a meadow, or a steppe is similar; perhaps slower, and the richness of the layer of humus will depend on a good many factors. One, in particular, has an obvious effect, namely, the supply of air. If, for some reason, this is cut off, the formation of humus is greatly impeded. Areas, therefore, that are partly or completely waterlogged will not form humus as the forest does: the upper portion of the soil will not have access to sufficient free oxygen, nor will there be much oxygen in the standing water. In the first case a moor will result; in the second a bog or morass will be formed. In both these the conditions are anaerobic: the organisms derive their oxygen not from the air but from the vegetable and animal residues including the proteins. In this fermentation nitrogen is always lost and the resulting low-quality humus is known as peat.

But the forest, the prairie, the moor, and the bog are not the only areas where humus formation is in progress. It is constantly going on in the most unlikely places—on exposed rock surfaces, on old walls, on the trunks and branches of trees, and indeed wherever the lower forms of plant life—algae, lichens, mosses, and liverworts—can live and then slowly build up a small store of humus.

Nature, in fact, conforming to that principle of reserves, does not attempt to create the higher forms of plant life until she has secured a good store of humus. Watch how the small bits of decayed vegetation fall into some crack in the rock and decompose: here is the little fern, the tiny flower, secure of its supply of food and well able to look after itself, as it thrusts its roots down into the rich pocket of nourishment. Nature adapts her flora very carefully to her varying supplies of humus. The plant above is the indicator of what the soil below is like, and a trained observer, sweeping his eye over the countryside, will be able to read it like the pages of a book and to tell without troubling to cross a valley exactly where the ground is waterlogged, where it is accumulating humus, where it is being eroded. He looks at the kind and type of plant, and infers from their species and condition the nature of the soil which they at once cover and reveal.

But we are not at the end of the mechanisms employed by Nature to get her great Wheel to revolve with smooth efficiency. The humus that lies on the surface must be distributed and made accessible to the roots of plants and especially to the absorbing portions of the roots and their tiny prolongations known as root hairs—for it is these which do the delicate work of absorption. How can this be done? Nature has, perforce, laid her accumulation on the surface of the soil. But she has no fork or spade: she cannot dig a trench and lay the food materials at the bottom where the plant root can strike down and get them. It seems an impasse, but the solution is again curiously simple and complete. Nature has her own labour force—ants, termites, and above all earthworms. These carry the humus down to the required deeper levels where the thrusting roots can have access to it. This distribution process goes on continually, varying in intensity with night and day, with wetness or dryness, heat or cold, which alternately brings the worms to the surface for fresh supplies or sends them down many feet. It is interesting to note how a little heap of leaves in the garden disappears in the course of a night or two when the earthworms are actively at work. The mechanism of humus distribution is a give and take, for where a root has died the earthworm or the termite will often follow the minute channel thus created a long way.

Actually the earthworm eats of the humus and of the soil and passes them through its body, leaving behind the casts which are really enriched earth—perfectly conditioned for the use of plants. Analyses of these casts show that they are some 40 per cent richer in humus than the surface soil, but very much richer in such essential food materials as combined nitrogen, phosphate, and potash. Recent results obtained by Lunt and Jacobson of the Connecticut Experiment Station show that the casts of earthworms are five times richer in combined nitrogen, seven times richer in available phosphate, and eleven times richer in potash than the upper six inches of soil.

It is estimated that on each acre of fertile land no less than twenty-five tons of fresh worm casts are deposited each year. Besides this the dead bodies of the earthworms must make an appreciable contribution to the supply of manure. In these ways Nature in her farming has arranged that the earth itself shall be her manure factory.

As the humus is continually being created, so it is continually being used up. Not more than a certain depth accumulates on the surface, normally anything from a few inches to two or three feet. For after a time the process ceases to be additive and becomes simply continuous: the growing plants use up the product at a rate equalling the rate of manufacture—the even turning of the Wheel of Life—the perfect example of balanced manuring. A reserve, however, is at all times present, and on virgin and undisturbed land it may be very great indeed. This is an important asset in man's husbandry; we shall later see how important.

THE IMPORTANCE OF MINERALS

Is the humus the only source from which the plant draws its nourishment? That is not so. The subsoil, i.e. that part of the soil derived from the decay of rocks, which lies below the layer of humus, also has its part to play. The subsoil is, as it were, a depository of raw material. It may be of many types, clay, sand, etc.; the geological formation will vary widely. It always includes a mineral content—potash, phosphates, and many rarer elements.

Now these minerals play an important part in the life of living things. They have to be conveyed to us in our food in an organic form, and it is from the plant, which transforms them into an organic phase and holds them thus, that we and the other animals derive them for our well-being.

How does the plant obtain them? We have seen that there is a power in the roots of all plants, even the tiniest, of absorbing them from the soil solution. But how is the soil solution itself impregnated with these substances? Mainly through the dissolving power of the soil water, which contains carbon dioxide in solution and so acts as a weak solvent. It would appear that the roots of trees, which thrust down into the subsoil, draw on the dissolved mineral wealth there stored and absorb this wealth into their structure. In tapping the lower levels of water present in the subsoil—for trees are like great pumps drawing at a deep well—they also tap the minerals dissolved therein. These minerals are then passed into all parts of the tree, including the foliage. When in the autumn the foliage decays and falls, the stored minerals, now in an organic phase, are dropped too and become available on the top layers of the soil: they become incorporated in the humus. This explains the importance of the leaf-fall in preserving the land in good heart and incidentally is one reason why gardeners love to accumulate leaf-mould. By this means they feed their vegetables, fruit, and flowers with the minerals they need.

The tree has acted as a great circulatory system, and its importance in this direction is to be stressed. The destruction of trees and forests is therefore most injurious to the land, for not only are the physical effects harmful—the anchoring roots and the sheltering leaf canopy being alike removed—but the necessary circulation of minerals is put out of action. It is at least possible that the present mineral poverty of certain tracts of the earth's surface, e.g. on the South African veldt, is due to the destruction over wide areas and for long periods of all forest growth, both by the wasteful practices of indigenous tribes and latterly sometimes by exploiting Western interests.

SUMMARY

Before we turn to consider the ways in which man has delved and dug into all these riches and disturbed them for his own benefit, let us sum up with one final glance at the operations of Nature. Perhaps one fact will strike us as symptomatic of what we have been reviewing, namely, the enormous care bestowed by Nature on the processes both of destruction and of storage. She is as minute and careful, as generous in her intentions, and as lavish in breaking down what she has created as she was originally in building it up. The subsoil is called upon for some of its water and minerals, the leaf has to decay and fall, the twig is snapped by the wind, the very stem of the tree must break, lie, and gradually be eaten away by minute vegetable or animal agents; these in turn die, their bodies are acted on by quite invisible fungi and bacteria; these also die, they are added to all the other wastes, and the earthworm or ant begins to carry this accumulated reserve of all earthly decay away. This accumulated reserve—humus—is the very beginning of vegetable life and therefore of animal life and of our own being. Such care, such intricate arrangements are surely worth studying, as they are the basis of all Nature's farming and can be summed up in a phrase—the Law of Return.

We have thus seen that one of the outstanding features of Nature's farming is the care devoted to the manufacture of humus and to the building up of a reserve. What does she do to control such things as insect, fungous, and virus diseases in plants and the various afflictions of her animal kingdom? What provision is to be found for plant protection or for checking the diseases of animals? How is the work of mycologists, entomologists, and veterinarians done by Mother Earth? Is there any special method of dealing with diseased material such as destruction by fire? For many years I have diligently searched for some answer to these questions, or for some light on these matters. My quest has produced only negative evidence. There appears to be no special natural provision for controlling pests, for the destruction of diseased material, or for protecting plants and animals against infection. All manner of pests and diseases can be found here and there in any wood or forest; the disease-infected wastes find their way into the litter and are duly converted into humus. Methods designed for the protection of plants and animals against infection do not appear to have been provided. It would seem that the provision of humus is all that Nature needs to protect her vegetation; and, nourished by the food thus grown, in due course the animals look after themselves.

In their survey of world agriculture—past and present—the various schools of agricultural science might be expected to include these operations of Nature in their teaching. But when we examine the syllabuses of these schools, we find hardly any references to this subject and nothing whatever about the great Law of Return. The great principle underlying Nature's farming has been ignored. Nay more, it has been flouted and the cheapest method of transferring the reserves of humus (left by the prairie and the forest) to the profit and loss account of *homo sapiens* has been stressed instead. Surely there must be something wrong somewhere with our agricultural education.

3

SYSTEMS OF AGRICULTURE

WHAT IS AGRICULTURE? It is undoubtedly the oldest of the great arts; its beginnings are lost in the mists of man's earliest days. Moreover, it is the foundation of settled life and therefore of all true civilization, for until man had learnt to add the cultivation of plants to his knowledge of hunting and fishing, he could not emerge from his savage existence. This is no mere surmise: observation of surviving primitive tribes, still in the hunting and fishing stage like the Bushmen and Hottentots of Africa, show them unable to progress because they have not mastered and developed the principle of cultivation of the soil.

PRIMITIVE FORMS OF AGRICULTURE

The earliest forms of agriculture were simple processes of gathering or reaping. Man waited until Nature had perfected the fruits of the earth and then seized them for his own use. It is to be noted that what is intercepted is often some form of Nature's storage of reserves; more especially are most ripe seeds the perfect arsenals of natural reserves. Interception may, however, take other forms. A well-developed example of human existence based on a technique of interception is the nomadic pastoral tribe. Pastoral peoples are found all over the world; they have played some part in the history of the human race and often exhibit an advanced degree of culture in certain limited directions, not only material. Their physical existence is sustained on what their flocks and herds produce. To secure adequate grazing for their animals they wander, sometimes to and fro between recognized summer and winter pastures, sometimes over still greater distances. In this way they intercept the fresh vegetable growths brought to birth season by season out of the living earth; however successful, it is nothing more than a harvesting process.

It is presumed rather than known that at some period man extended his idea of harvesting to the gathering of the heads of certain plants, thus adding a vegetable element to the milk, meat, and fish he had been deriving from his animals and the chase. Wild barley, rice, and wheat are all supposed to have been gathered in this way in different parts of the earth. But real agriculture only began when, observing the phenomenon of the germination of seeds, instead of consuming all that they had gathered, men began to save some part of what they had in store for sowing in the ground. This forced them to settlement, for they had to wait until the plants grew from the seed and matured.

If at first the small store of gathered seed was sown in any bare and handy patch, the convenience of clearing away forest growths so as to extend the space for sowing soon became apparent. The next stage was to prepare the ground thus won. The art of tillage has progressed over the centuries. The use of a pointed stick drawn through the ground is still quite common. The first ploughs were drawn by human labour—a practice which survived even in such countries as Hungary and Romania into the nineteenth century. But the use of animals, tamed for their muscular strength to replace the human team, became the normal and world-wide practice, until ousted in certain continents first by the still more powerful steam engine and now by the internal combustion engine.

What was the purpose of this tillage, which is still the prime agricultural process? The first effect is, of course, physical. The loosened soil makes room for the seed, which thus can grow in abundance, while to cover the sowing with scattered earth or to press it into the ground protects it from the ravages of birds or insects. Secondly, tillage gives access to the air—and the process of soil respiration starts up, followed by the nitrification of organic matter and the production of soluble nitrates. The rain, too, can penetrate better. In this way physical, biological, and chemical effects are set in motion and a series of lively physiological changes and transformations result from the partnership between soil and plant. The soil produces food materials: the plants begin to grow: the harvest is assured: the sowing has become a crop.

Yet this is not the way in which Nature is accustomed to work. She does not, as a rule, collect her plants, the same plants, in one spot and practise monoculture, but scatters them: her mechanisms for scattering seed are marvellous and most effective. Man's habit, so convenient, of collecting a specified seed and sowing it in a specified area implies, it must be acknowledged, a definite interference with Nature's habits. Moreover, by consuming the harvest and thus removing it from the place where it had grown he for the time being interrupts the round of natural processes.

In fact, man has laid his hand on the great Wheel and for a moment has stopped or deflected its turning. To put it in another way, he has for his own use withdrawn from the soil the products of its fertility. That man is entitled to put his hand on the Wheel has never been doubted, except by such sects as the Doukhobors who argued themselves into a state of declaring it a sin to wound the earth with spades or tools. But if he is to continue to exist, he must send the Wheel forward again on its revolutions. This is a necessary part of all primitive cultivation practices and perhaps a tenet of all true early religions as soon as they lift themselves from the stages of mere animism or fetish worship; at any rate, all the great agricultural systems which have survived have made it their business never to deplete the earth of its fertility without at the same time beginning the process of restoration. This becomes a veritable preoccupation.

SHIFTING CULTIVATION

The simplest way of doing this is after a time to leave the cultivated patch and thus stop the process of interference. Nature will overrun it again with scrub or forest: soon the green carpet is re-established: in due course humus will accumulate: it will be as it was—the earth's fruitfulness will be restored. To pass on, therefore, from one patch to another, and again to another and another, is a common primitive practice found in Africa, India, Ceylon, and many other parts of the world, and is known as shifting cultivation. It even occurred in the American continent some ten years or so ago before the Tennessee Valley Authority was constituted by the late President of the United States of America. In this shifting cultivation the fresh patch is usually cleared by burning the jungle: this leaves the ash *in situ,* and thus retains some of the mineral contents of the burnt vegetation for the benefit of the coming crop. But it is a wasteful method, for a large aggregate area is required to feed a small group, while a long period has to be reckoned to replace the lost fertility. Indeed, this replacement is seldom consummated. The larger trees suffer, the best part of the forest is virtually destroyed. It will also be observed that after using up the riches of the soil man actually does nothing to restore it—he merely leaves it. This lazy practice constitutes the least satisfactory of many agricultural systems and,

35

entailing constant small movements of working area on the part of those practising it, is no foundation for a settled civilization. It does, however, show that primitive tribes not only realized the fact that fertility can be exhausted, but also understood how it could be restored.

THE HARNESSING OF THE NILE

A much more satisfactory method of restoring soil fertility was evolved in the great river valley of the Nile which, according to some theorists, was the original home of agriculture proper. It is the peculiarity of this great river that it overflows once a year with great regularity, bearing suspended in its flood an accumulation of fertile silt washed down from its catchment basin; this accumulation, rich in both mineral and organic matter, is gently deposited and is capable of yielding an abundant harvest. The process continued for centuries. Early engineering skill led the silt-laden water to embanked fields by means of inundation canals. The deposit was trapped just where it was needed and the land was at the same time saturated with water. When the embanked fields were dry enough, they were ploughed and sown: no rain fell and no more water was needed for a full crop. The annual additions of rich silt made this method of farming permanent. In this way there grew up settled habitations, a great civilization, an historic people.[1]

STAIRCASE CULTIVATION

Few areas on the earth's surface are so fortunate. What the great river bestowed on the lucky Egyptians has had to be created in other parts of the world, sometimes in the most unpromising conditions. The so-called staircase cultivation of the ancient Peruvians is regarded as one of the

[1] This basin system of irrigation in Egypt, which is perhaps the best and most permanent that can be devised, has of recent years been replaced by another—perennial irrigation—by which the same field can be watered periodically to allow of cotton being grown. For this purpose the Nile has been impounded and a vast reservoir has been created for feeding the canals. But unless the very greatest care is taken to restore and then to maintain the compound soil particles by means of constant dressings of freshly prepared humus these modern methods are doomed. The too frequent flooding of the close silts of this river valley will lead to the formation of alkali salts and then to the death of the soil. This will be the fate of Egypt if the powers-that-be persist in the present methods of cultivation of cotton and do not realize before it is too late that their ancient system of irrigation is, after all, the best. Will a few years of cotton growing make up for the loss of the soil on which the very life of Egypt is based? On the answer to this question the future of the Nile valley will depend.

oldest forms of agriculture known to us—it dates from the Stone Age. Without metal tools this people could not remove the dense forest growths of the humid South American valleys. They were driven to the upland areas under grass, scrub, or stone. Here they constructed terraced fields up the slopes of the mountains, tier upon tier, sometimes as many as fifty tiers rising one above the other. The outer retaining walls of these terraces were made of large stones fitted into each other with such accuracy that even at the present day a knife blade cannot be inserted between them. Inside these walls were laid coarse stones and over these clay, then layers of soil several feet thick, all of which had to be imported from beyond the mountains. Just sufficient slope was given to each tiny field for watering, water also being brought in stone aqueducts from immense distances—one aqueduct of between 400 and 500 miles has been found traversing the mountain slope many hundreds of feet above the valley. Thus a series of gigantic flower pots were formed and in these were grown the crops to nourish a nation and to establish a civilization.

The results of such incredible labour are still to be seen, but the Inca nation itself has vanished. However, in the Hunzas living in a high mountain valley of the Gilgit Agency on the Indian frontier we have an existing demonstration of what a primitive system of agriculture can do if the basic laws of Nature are faithfully followed. The Hunzas are described as far surpassing in health and strength the inhabitants of most other countries; a Hunza can walk across the mountains to Gilgit sixty miles away, transact his business, and return forthwith without feeling unduly fatigued. In a later chapter we shall point to this as illustrative of the vital connection between a sound agriculture and good health. The Hunzas have no great area from which to feed themselves, but for thousands of years they have evolved a system of farming which is perfect. Like the ancient Peruvians they have built stone terraces, whose construction admits of admirable soil drainage and therefore of admirable soil aeration—for where water drains away properly air is abundantly drawn in. As in the ancient Peruvian system, irrigation is employed to obtain the water and it is not without interest that this water is glacier water bringing down continual additions of fine silt ground out from the rocks by the great cap of ice. It is probable, though it has not been investigated, that the mineral requirements of the fields are thus replenished to a remarkable degree. To provide the essential humus every kind of waste, vegetable, animal, and human, is mixed and decayed together by the cultivators and incorporated into the soil;

the law of return is obeyed, the unseen part of the revolution of the great Wheel is faithfully accomplished.

THE AGRICULTURE OF CHINA

It is this return of all wastes to the soil, including the mud of ponds, canals, and ditches, which is the secret of the successful agriculture of the Chinese. The startling thing to realize about this peasant nation of over four hundred million souls is the immense period of time over which they have continued to cultivate their fields and *keep them fertile,* at least 4,000 years. This is indeed a contrast to the shifting cultivation of the African and it may be observed here that the greatest misfortune of the African continent has been that it never came into contact with the agricultural peoples of the Far East and never revised its systems of cultivation in the light of the knowledge it might thereby have gained— the great lesson of the Nile basin was not truly apprehended and has had no influence outside Egypt, whereas over large parts of eastern Asia the central problem of agriculture was solved very early, empirically and not by a process of scientific investigation, yet with outstanding success.

The Chinese peasant has hit on a way of supplying his fields with humus by the device of making compost. Compost is the name given to the result of any system of mixing and decaying natural wastes in a heap or pit so as to obtain a product resembling what the forest makes on its floor: this product is then put on the fields and is rich in humus. The Chinese pay great attention to the making of their compost. Every twig, every dead leaf, every unused stalk is gathered up and every bit of animal excreta and the urine, together with all the wastes of the human population, are incorporated. The device of a compost heap is clever. By treating this part of the revolution of the Wheel as a special process, separated from the details of cultivation, time is gained, for the wastes mixed in a heap and kept to the right degree of moisture decay very quickly, and successive dressings can be put on the soil, which thus is kept fed with just what it needs: there is no pause while the soil itself manufactures from the raw wastes the finished humus. On the contrary, everything being ready and the humus being regularly renewed at frequent intervals, the soil is able to feed an uninterrupted succession of plants, and it is a feature of Chinese cultivation that one crop follows another without a pause, indeed crops usually overlap, the ripe crop being skilfully removed by hand from among the young growing plants of the succeeding planting or sowing. In short, what the Chinese farmer really does is ingeniously to extend his area. He, so to say, rolls up the floor of the forest and arranges it in a heap. The great processes of decay go on throughout that heap, spreading themselves over the whole of the internal surface of the heap, that is, over the whole of the surfaces implied in the juxtaposition of every piece of waste against every other. He also overcomes the smallness of the superficial area of his holding by increasing the internal surface of the pore spaces of his soil. This is what matters from the point of view of the crop—the maximum possible area on which the root hairs can collect water and food materials for the green leaf. To establish and to maintain this maximum pore space there must be abundant humus, as well as a large and active soil population.

Thus is created the most intensive agriculture which the world has so far seen. Each Chinese family lives on the produce of a very tiny piece of ground, an area which would mean downright starvation in most other countries. In spite of great calamities which repeat themselves, principally floods, the causes of which will be mentioned hereafter, the Chinese peasant may be said to be, on the whole, well nourished. His resisting power to the many frightful diseases, sufficient to kill off most other populations, has been noted, while the standard of culture which he has reached and has maintained over the long period of his existence rivals the contributions of Western civilization.

He is indeed the classic example of a nation which has conserved the fertility of its soil. Other nations have done the same, but none over so long a period or on so vast an area. Is it legitimate to interpret the history of the nations by the way in which they have made use of the land which chance or their own valour assigned to them? We have considered some instances where attempts have been made to conserve fertility with greater or lesser success. Let us now turn to some different examples.

THE AGRICULTURE OF GREECE AND ROME

The agricultural history of the ancient Greeks is not altogether clear. But one thing is certain: in common with most other Mediterranean peoples they permitted an extraordinary amount of destruction of forest growths over some of the areas bordering on this great inland sea.

Greece is now a land bare of trees and the continued depredations of the goat have done untold harm to any young growths that have attempted to survive. Whether this process began on a large scale very early and whether the result was a severe disturbance of the drainage of a not very fruitful country, extending on the one hand the area of marsh and on the other inviting erosion, is not certain. Such conditions would affect first the crops and then those who fed off them—subtle forms of undernourishment and disease would appear. The theory has been put forward that the extraordinary and unexplained collapse of the Greek nation in the fourth and third centuries B.C., after a period of the highest vigour and culture, was due to the spread of malaria. It is a theory which is very reasonable and would explain much.

The case of the Romans, another Mediterranean people, is not quite the same. For many centuries they maintained a flourishing agriculture to which they paid great attention. The backbone of the nation throughout its greatest period was the staunch mass of smallholders, each engaged on cultivating his own farm and only breaking off at intervals to pursue political matters with great vigour or to fight short summer campaigns with the utmost zeal. In spite of the attractions of the metropolis and of the wonderful educational influence with which city life shaped law, thought, and conduct, the rural background was conserved and valued; religion remained rather rural throughout and never got very much beyond the peasant outlook. It was the necessity for fighting prolonged foreign compaigns which destroyed all this. Then came the fatal attractions of slave labour. The smallholder was tempted or indeed was obliged to desert his holding for years. Such holdings began to be bought up, for wealth accumulated from the spoils of the East. Slaves were drafted in to work these agglomerations of great estates: the evil *latifundium,* which means the plantation in its worst form, spread everywhere. The final phase was reached when tillage was given up for the cheaper pastoral industry: where there had been countless flourishing homesteads now ranged great herds of cattle tended by a few nomadic shepherd slaves.

This disastrous change, which was deeply deplored by such writers as Cicero, lasted and, except in northern Italy, was not made good. A few years ago it was possible to see on a mere day's excursion away from Rome a wild shepherd tending his sheep over a ruined countryside which might have been carved out of the most ancient of wildernesses, so entirely was it denuded of all traces of tillage or of the care of man.

There must have been some profound upsetting of the balanced processes of Nature to reduce so fertile a country as Italy to such a state and Nature in revenge has preferred to continue her revolution of the Wheel on the lowest gear, spreading her marsh, her scrub, and her desert, where once there were fields and meadows.

Having largely destroyed the food-bearing capacity of the Italian peninsula, the Romans were forced to feed their swollen cities from elsewhere. For the dispossessed rural population drifted to the towns, which became further congested with a great influx of foreigners and foreign slaves: all had to be fed, and Alexandria and Antioch were problems no less great than Rome. First Sicily and then North Africa, at that time great wheat-growing countries, were exhausted. We cannot trace the process and do not know how much to attribute to a false economy, how much to the ravages of centuries of war, as wave after wave of conquerors disputed possession. When these countries reappear after such cataclysms, Sicily is a wild pastoral country, North Africa, except for a few coastal tracts and, of course, always Egypt, a desert.

FARMING IN THE MIDDLE AGES

The rest of the continent of Europe was more fortunate. Out of the lingering shadows of the Roman Empire there finally emerged into medieval times a system of agriculture which held its own well into the nineteenth century. Such a long history is an honourable one and we may agree that this system, that of mixed husbandry, was in many essentials excellent. Except where a frozen legal system ground down the cultivator—"trembling peasants gathering piteous harvests"—both the large farm and the smallholding, the landlord and the tenant, survived in good health and considerable comfort. Food was abundant and nourishing, and above all the soil remained in good heart.

The system depended on certain principles. In the first place, animal husbandry was practised alongside of the production of vegetable crops: there was thus a supply of manure. The manure was not made on the most perfect system. The European manure heap, normally regarded as the inevitable method of collecting and storing animal wastes, is nevertheless most inefficient, as will be pointed out in a later chapter (p. 204). But it has played a prime role in maintaining the fertility of the continent, although it is wasteful and extravagant, unhealthy, and unnatural:

41

with the help of the manure heap the return of much of the wastes of farming was assured to the land.

The use of the cesspit was even less successful and it is not surprising that water-borne sewage, when once invented, rapidly replaced it: unfortunately this permitted the final escape of valuable wastes to the sea. To this came to be added, also in the course of the nineteenth century, the further loss of all dustbin refuse which, again on the dictates of the new sanitary science, was destroyed by burning or was buried in unused tips. Nevertheless, until these modern sewage disposal methods were developed, it is significant that all material wastes went back to the soil in however imperfect a way.

A third principle in conserving fertility was the fallow. Arable land was rested by allowing it to remain idle for a year or for a longer period by the establishment of a temporary carpet of grass and weeds. A part at least of the advantage of the bare fallow was the benefit conferred by the weeds. When laid down to grass for sheep, the green carpet rapidly deposited a mass of vegetable wastes under the turf which, with the turf and the animal wastes deposited thereon, provided all the raw materials for sheet-composting when the land came under the plough. Both these methods have been employed in European farming for many centuries and did much to conserve the fertility of the soil.

As long as all these principles governed European farming it could roughly hold its own, although a slow running down of soil fertility remained at all times a possibility, as will be seen in the next chapter. It began to break down seriously with the advent of the Industrial Revolution. But before dealing with the changes thus brought about in European agriculture it will be illuminating to examine in greater detail the story of one people, our own, in terms of the use made by the community of soil fertility. We shall see that, in spite of the great and advantageous practices to which we have alluded, soil fertility was subtly and gradually used up. This has determined much in our national affairs.

4

THE MAINTENANCE OF SOIL
FERTILITY IN GREAT BRITAIN

Many accounts of the way the present system of farming in Great Britain has arisen have been published. The main facts in its evolution from Saxon times to the present day are well known. Nevertheless, in one important respect these surveys are incomplete. Nowhere has any attempt been made to bring out the soil fertility aspect of this history and to show what has happened all down the centuries to that factor in crop production and animal husbandry—the humus content of the soil—on which so much depends. The present chapter should be regarded as an attempt to make good this omission.

THE ROMAN OCCUPATION

At the time of the Roman invasion most of the island in which we are living was under forest or marsh: only a portion of the uplands was under grass or crops: the population was very small. After the conquest of the country the Romans began to develop it by the creation on the areas already cleared of an agricultural unit—new to Great Britain—known as the villa. These villas were large farms under single ownership run by functionaries each responsible for a particular type of animal or crop and worked by slave labour. These units followed to some extent the methods of the *latifundia* of Italy and were designed for the production of food for the legions garrisoning the island and those stationed in Gaul. Wheat—an exhausting crop—was an important item in Roman agriculture, for the reason that this cereal provided the chief food (*frumentum*) of the soldiers. The extent of the export of grain to Gaul will be evident from the fact that in the reign of the Emperor Julian no less than 800 wheat ships were sent from Britain to the Continent.

The exhaustion of the soils of the island began even before the Roman occupation. The heavy soil-inverting mould board plough, which invariably wears out the land, was already in use when the

43

Romans arrived, and was probably brought by the Belgic tribes who conquered and settled in the south-eastern part of the country. They lived in farmsteads and cultivated large open fields. They were highly skilled agriculturists and exported to Gaul a considerable quantity of their main product—wheat. This practice was developed by the Roman villas which followed and in this way the slow exhaustion of the lighter soils of the down lands of the south-east became inevitable.

After an occupation which lasted some 400 years and which contributed little or nothing of permanent value to the agriculture of the island beyond some well-designed roads, the legions evacuated the island and left the Romanized population to look after itself. This they failed to do: the country was soon conquered by the Saxon invaders, in the course of which much destruction of life and property took place. One result was the creation of a new type of farming.

THE SAXON CONQUEST

The settlement of Nordic people in our island is the governing event both of British history and of British agriculture. The new settlers had inhabited the belts of land around the Weser and the Elbe and their first contact with Britain was as raiders; their operations were in the nature of reconnaissance to ascertain the chances of settlement. The Anglo-Saxon migration to Britain was a colonization preceded by conquest, in which the farming system of the Romanized population was, in the midland area at any rate, destroyed. In the east, south-east, and western portions of the island some relics of Roman and Celtic methods survived.

Our forefathers brought with them from the opposite shores of the North Sea their wives, children, livestock, and a complete fabric of village life. The immigrants, being country folk, wanted to live in rural huts with their cattle round them and their land nearby, as they did in Germany. The numerous villages they formed reproduced in all essentials those they had left behind on the mainland. Our true English villages are, therefore, not Celtic, not Roman, but purely and typically German.

The Roman villas were replaced by a new system of farming—the Saxon manor—in which the tenants held land in return for service. The lord and his retainers shared the land, each bound to perform

44

certain duties determined by custom. The manors took centuries to evolve. By A.D. 800 they had developed into a permanent system which provided the material for the Domesday Book of the Normans, by which taxation was assessed and a rigid feudal system became firmly established.

THE OPEN-FIELD SYSTEM

The first general feature that strikes us in early Anglo-Saxon England is the strip cultivation of the arable land on the open-field system. This system was a communal agricultural institution started by people who had to get a living out of the soil. They had progressed so far as to use the plough and had a common fund of experience. Everyone pursued the same system of farming. The arrangement of the open fields was, however, by no means uniform. No fewer than three distinct types arose, corresponding to as many different influences exerted by people who had early occupied the country. The large central midland area, stretching from Durham to the Channel and from Cambridgeshire to Wales, is the region where Germanic usage prevailed. The south-east was characterized by the persistence of Roman influence, a circumstance which implies that the conquest was less destructive there than in the north and west. The counties of the south-west, north-west, and the north retained Celtic agrarian usages in one form or another, which is easily understood in view of the difficulty with which, as we know, these districts were slowly overpowered by the invaders. The midland area was thus the region where the Anglo-Saxons were most firmly established and where the subjugation of the fifth century was most thorough. The Romano-Celtic people who remained were not numerous enough to preserve any traces of Roman or Celtic methods of tilling the soil.

Throughout this extensive region a two-field and a three-field system, or sometimes a mixture of the two, prevailed. This field arrangement was a custom prevalent in Germany, especially east and south of the Weser. The chief characteristic of the two- and three-field type of tillage was the distribution of the parcels of arable land (which made up the holdings of the customary tenants) equally amongst the two or three fields. The cropping was so arranged that one field in the two-field system and two fields in the three-field system were cropped every year, and thus one-half or one-third of the township's arable land lay fallow

45

and was used for common grazing—a point which is always emphasized in the midland system.

Besides the cultivated open fields, for which the best land was always used, the village lands consisted of grassland for mowing on the wetter parts, and commons or woodlands on the poorer parts.

Ploughing was the all-important operation of medieval tillage and was carried out on a co-operative basis, and demanded a team of eight draught animals yoked to a heavy plough. This, of course, was beyond the reach of any but the largest and most prosperous tenants. Communal ploughing in Saxon times was, therefore, inevitable. It was the difficulty of replacing this communal ploughing that delayed agricultural progress in many parts of the country.

The open-field system repeated itself for centuries, not only in England but in a great part of Europe—nations living under very different conditions, in very different climates, and on very different soils adopted the open-field system again and again without having borrowed it from each other. This could not but proceed from some pressing necessity. The open-field system is communal in its very essence. Every trait which makes it strange and inconvenient from the point of view of individualistic interests renders it highly appropriate to a state of things ruled by communal conceptions—right of common usage—communal arrangements of ways and time of cultivation. These are the main features of open-field husbandry and all points to one origin—the formation in early Anglo-Saxon society of a village community of shareholders of free and independent growth.

It must be borne in mind that the open-field prevailed during the period of national formation of the English people and its influence on the life of the village community must have been very great. The sense of personal responsibility, which the system of communal work created, made it a vital factor in the social education of the people.

THE DEPRECIATION OF SOIL FERTILITY

Open-field farming is, as a rule, balanced: the fertility used up in growth is made good before the next crop is sown. Compared with our modern standards, however, the yield is remarkably low and the removal of fertility by such small crops is made up for by the recuperative processes operating in the soil (non-symbiotic fixation of nitrogen and so forth). The surplus of available humus originally left by the forest is depleted at an early stage and an equilibrium is established, the yield adjusting itself to the amount of fertility added each year by natural processes, this in its turn is influenced by climate and methods of cultivation.

For example, in the peasant cultivation of north-west India at the present day a perfect balance has been established between losses and gains of fertility. The village land on which corn crops are grown has been cultivated for upwards of 2,000 years without manure beyond the droppings of the livestock during the fallow period between harvest and the rains. But the Indian cultivators use primitive scratch ploughs and are most careful not to draw on the reserves of organic material in the soil, as its texture depends on this. They produce crops entirely on the current account provided by the annual increments of fertility. The yield has settled down to 8 maunds (658 lb. per acre) of wheat on unirrigated land, and 12 maunds (987 lb.) of wheat on irrigated land, and this yield has been constant for many centuries.

The same processes were operating in the English open fields. The reserve of humus in the soils originally under forest, which the Saxons brought into cultivation, was soon used up and the yield was determined by the annual additions of fertility to the soil by natural means. But in our cold and sunless climate and on our ill-drained, poorly aerated soils this is far less than in the semi-tropical conditions of northern India. Moreover, and this point must be stressed, the Saxons from the earliest times used a soil-inverting plough, which has a marked tendency to exhaust the humus in the soil if provision is not made for the regular supply of sufficient farmyard manure. In fact, recent experience in many parts of the world is proving that the continued use of heavy soil-inverting, tractor-driven implements, without sufficient farmyard manure to manure the land, promptly leads to catastrophic consequences.

The first recorded references to the mould board plough speak of it in Gaul, but some authorities quoted by Vinogradoff (*The Growth of the Manor*) suggest that it was borrowed by the Germanic people from the Slavs, and in view of the soil types found in Slav territory this may easily be so. The evolution of the big plough was due to soil requirements as settled agricultural life developed in the heavy, moist soils of north Europe after the forests had been cleared.

The mould board plough determined the lay-out of the open fields It divided the arable areas into a succession of lands. It needed a headland to turn on, and there was a limit to the length of furrow a team of oxen could plough before needing the relief got by stopping and turning. This furrow-long or furlong became one of our units of length. It was usual to keep the land in high ridges running along the slopes to facilitate surface drainage, an important point in England. The ridges varied in width according to the nature of the soil. In very heavy clays they were sometimes no more than three yards wide. In lighter soils they might be twenty-two yards wide. These ridges may be seen in many places to-day on grassland which was under the plough in earlier centuries. From this brief description it will be seen that the open fields cultivated with the heavy medieval plough were laid out in strips.

The main feature of the heavy mould board plough was its high penetrating power, and it could be used on the heavier types of soil where the light scratch plough of the Celts and Italians would be useless. It thus enabled the cropped area in England to be greatly extended by the cultivation of the heavy soil of the valleys and plains which first had to be slowly carved out of the forest. It owed its superiority to an iron share, a coulter, and a wooden mould board so suitable on wet land. This primitive implement gave us the plough as we know it to-day. The principle of our modern plough is identical and, except for the fact that it is now made entirely of iron, it is almost the same in detail.

The open-field system of the Middle Ages was bound to fail because it involved burning the candle at both ends and also in the middle. First the natural recuperation processes in the soil were hampered by low temperatures and poor soil aeration; second, such supplies of farmyard manure as were available were by custom mostly bestowed on the lord's demesne lands, and besides were inadequate because only a portion of the livestock could be wintered; finally the soil-inverting plough led to the oxidation of the stores of soil humus faster than it could be recreated and was bound to wear out the land.

THE LOW YIELD OF WHEAT

The failure of the open-field system is proved by the low yield of wheat. All authorities agree that the yield of wheat in England during the Middle Ages was at a very low level, though it does not appear to have varied greatly. It may be noted that there was never any question of complete exhaustion of the wheat-growing land, such as occurred in Mesopotamia and in the Roman wheat-growing regions of North

Africa, where the soil, owing to over-cropping and in some instances to over-irrigation aggravated by special climatic conditions, became sterile and was transformed into desert. This could not so easily happen in the moist, temperate climate of Great Britain. What happened in the Middle Ages in England was that the yield of corn was not high enough for the requirements of the growing social and economic life of the country.

The material for a quantitative estimate of wheat yields in this period is necessarily very scanty, but in the case of some large estates records are available for a considerable period of years of the seed sown in one year and the grain threshed in the following year, and these form the basis of the best estimates of medieval yields. Sir William Beveridge (*Economic Journal Supplement*, May 1927), using this method, investigated the yield of wheat for the years 1200 to 1450 on eight manors, including that of Wargrave, situated in seven different counties belonging to the Bishop of Winchester. The average yield per acre was 1.17 quarters or 9.36 measured bushels, equivalent to 7.48 bushels of 60 lb. It is to be noted that these estimates were all from demesne lands which were probably better cultivated and better manured than the land of the customary tenants. Other authorities confirm these figures.

The figures of yield given above help to account for the changes which marked the end of the Middle Ages. The amount of food was becoming insufficient for the growing population. But another factor was steadily developing, which finally assumed the dimensions of an avalanche and led to the reform of manorial farming. This was disease, a matter which must now be discussed.

THE BLACK DEATH

That the agriculture of the Middle Ages was unable to keep the population in health was first indicated by the frequent indications of rural unrest. But these were soon followed by the writing on the wall in the shape of the Black Death in 1348–9. This outbreak had been preceded by several years of dearth and pestilence, and it was succeeded by four visitations of similar disease before the end of the century. During its ravages it destroyed from one-third to one-half of the population. This seriously affected the labour supply, which was no longer sufficient to carry on the traditional methods of manorial farming, already beginning to be undermined by the growing tendency to replace service by money payments.

Land which could no longer be ploughed had to be laid down to grass and used for feeding sheep to produce more of the wool so urgently needed in Flanders and Lombardy. For the new farming the countryside had to be enclosed: first the lord's demesne and then the area under open fields began to be laid down to grass. The earth's green carpet not only fed the sheep, but gave the land a long rest: large reserves of humus were gradually built up under the turf: the fertility of the soil, which had been imperceptibly worn out by the mould board plough and the constant cropping of the manorial system, was gradually restored.

After a long period of rest of a century the land no longer returned only seven and a half bushels to the acre. The figures given above for the years 1200 to 1450 may be contrasted with the figures from a farm at Wargrave from 1612–20: in these years the average was 25.6 bushels of 60 lb. per acre (Beveridge, loc. cit.). In the latter part of the sixteenth century the general average was eighteen bushels to the acre and even more. That this significant change was due to the restoration of soil fertility by humus formation under the turf there can be no doubt.

It is more than probable that the slow regeneration of the soils of this country, which began after the Black Death, produced other results besides the improvement of crops and livestock. What of the effect of the produce of land in good heart on the most important crop of all—men and women? Were the outstanding achievements of the Tudor period one of the natural consequences of a restored agriculture? It may well be so.

ENCLOSURE

When increasing population led once more to the breaking up of the grassland and the farmer returned to tillage, the land, after its long rest of upwards of a century, was again capable of responding to the demands made upon it. One result of this experience was an increased interest in enclosure. Instinct was leading to a search for an economic arrangement which would prevent soil exhaustion from being repeated in succeeding ages. Enclosed farms offered a solution, as they gave the farmer the chance of keeping his land in good condition by individual management in place of the easy-going farming of the open fields of old English village agriculture. They also offered to the enclosed farmer the opportunity of composting his straw in his cattle yards and producing as much farmyard manure as possible. This, in most cases, he did, and the plan succeeded.

Nevertheless, the ancient open-field tillage husbandry had had in its favour the authority of long tradition—a potent force with a suspicious and conservative peasantry. The peasant asked himself: In the case of a readjustment of holdings would not the strong profit and the weak suffer? There grew up a popular prejudice against enclosure and the improvement of the common fields, but in the end, after some centuries of contest, enclosure won.

The form which the enclosure movement took before it was completed was due to the peculiar form of government which came in with the English Revolution of 1688. By that event the landed gentry became supreme. The national and local administration was entirely in their hands, and land, being the foundation of social and political influence, was eagerly sought by them. They not unnaturally wished to direct the enclosure movement into channels which were in the interests of their estates. But in doing so they made some of the most outstanding contributions to farming ever made in our history.

The restoration of soil fertility which resulted from enclosure had a profound influence on both livestock and crops. The provision of more and better forage and fodder which followed the cultivation of clover and artificial grasses, coupled with the popularization of the turnip crop by Townshend in 1730, opened the door for the continuous improvement of livestock by pioneers like Bakewell. The result was that our livestock improved in size and in the quality of the meat. Between 1710 and 1795 the weights of cattle sold at Smithfield more than doubled. By 1795 beeves weighed 800 lb. as compared with 370 lb.; sheep went up from 28 lb. to 80 lb. The improvement in the yield of cereals was no less significant. That of rye or wheat rose from 6–8 bushels to the acre in the Middle Ages to 15–20 bushels; barley yielded up to 36 bushels, oats 32–40 bushels. All this was due to more and better food for the livestock and more manure for the land. More manure raised larger crops: larger crops supported much bigger flocks and herds.

Another change in the countryside accompanied the enclosures. The forests, which since Saxon times had been gradually cleared and converted into manorial lands, had by this process become exhausted. After the Civil War it was realized that the country was running short of the hardwoods needed for maintaining the fleet and for buildings and so forth. An era of tree planting, which continued for two hundred years, was inaugurated by the publication of Evelyn's *Sylva* in 1678. It was during this period that the English landscape as we know it to-day was created by the judicious laying out of parks, artificial lakes, groups of trees, and woods. All this planting provided an important factor in the maintenance of soil fertility. The roots of the trees and the hedges combed the subsoil for minerals, embodied these in the fallen leaves and other wastes of the trees and shrubs, and so helped to maintain the humus in the soil, as well as the circulation of minerals. The roots also acted as subsoil ploughs and aerating agencies. The cumulative effect of the trees and hedges, which accompanied enclosure, in maintaining soil fertility has passed almost unnoticed. Nevertheless, its importance in humus production and in the availability of minerals must be considerable.

While the policy of enclosure, combined with tree-planting and the creation of the existing English landscape, arrested the fall in soil fertility which was inherent in the open-field system, the freedom of action which followed enclosure afforded full scope to the improver. The restoration of British agriculture owes much to the pioneers among the landlords themselves, particularly to Coke of Holkham (1776–1816), who did much to introduce the Norfolk four-course system—(1) turnips, (2) barley, (3) seeds (clover and rye grass), (4) wheat—into general practice and so to achieve at long last an approach to Nature's law of return. Besides his championship of the Norfolk four-course system, his achievements include the conversion of 2,000,000 acres of waste into well-farmed and productive land, the prevention of famine in England during the Napoleonic Wars, the solution of the rural labour problem in his locality by means of a fertile soil, the demonstration of the principle that money well laid out in land improvement is an excellent investment. He invested half a million sterling in his own property and thereby raised the rent roll of his estate from £2,200 a year to £20,000. He transformed agriculture in this country by the simple process of first writing his message on the land and then, by means of his famous sheep-shearing meetings, bringing it to the notice of the farming community.

But the replacement of the manorial system by individual farming in fenced fields was attended by some grave disadvantages. The large profits obtained from the sale of wool, for example, while they enriched the few, led to a new conception of agriculture. The profit motive began

PLATE III. LIVESTOCK FOR MAKING COMPOST ON A TEA ESTATE IN AFRICA

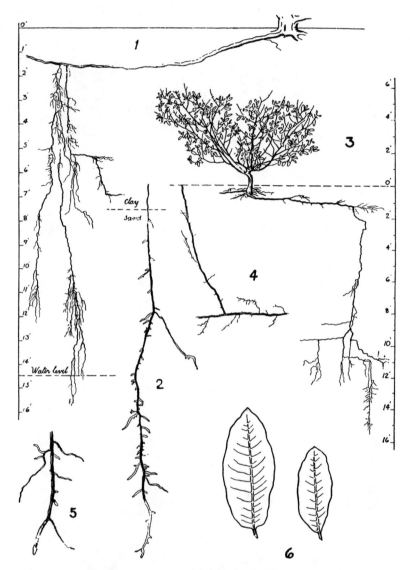

PLATE IV. GUAVA (*Psidium Guyava*, L.)

No. 1—Superficial and deep roots (November 23, 1921). No. 2—The
influence of soil texture on the formation of the rootlets (March 29,
1921). No. 3—The root-system under grass (April 21, 1921). No. 4—
Superficial rootlet growing to the surface (August 28, 1921). No. 5—
Formation of new rootlets in fine sand following the fall of the
ground water (November 20, 1921). No. 6—Reduction in the size
of leaves after twenty months under grass (right).

to rule the farmer; farming ceased to be a way of life and soon became a means of enrichment. Enterprising individuals were afforded considerable scope for using their farms to make money. At the same time, large numbers of less fortunate individuals deprived of their land had either to work for wages or seek a living in the towns.

The various Enclosure Acts, which covered a period of more than 600 years, 1235–1845, therefore led to a new agriculture, the enthronement of the profit motive in the national life, and to the exploitation of coal, iron, and minerals, which is customarily referred to as the Industrial Revolution. This arose from the activities of the tradesmen of the manor, whose calling was destroyed by the Enclosure Acts.

The last of the Enclosure Acts, which finally put an end to the strip system of the open fields, was passed in 1845. About the same time the celebrated Broadbalk wheat plots of the Rothamsted Experimental Station were laid out. This field is divided into permanent parallel strips and cultivated on even more rigid lines than anything to be found in the annals of manorial farming. These plots never receive the droppings of livestock: till recently they never had the benefit of the annual rest provided by a fallow. Practically every agricultural experiment station all over the world has copied Rothamsted and adopted the strip system of cultivation. How can such experiments, based on an obsolete method of farming, ever hope to give a safe lead to practice? How can the higher mathematics and the ablest statistician overcome such a fundamental blunder in the original planning of these trials?

The strip system has also been adopted for the allotments round our towns and cities without any provision whatsoever on the part of the authorities to maintain the land in good heart by such obvious and simple expedients as subsoiling, followed by a rest under grass grazed by sheep or cattle, ploughing up, and sheet-composting the vegetable residues. Land under allotments should not be under vegetables for more than five years at a time; this should be followed by a similar period under grass and livestock.

THE INDUSTRIAL REVOLUTION AND SOIL FERTILITY

The released initiative which accompanied the collapse of the manorial system was by no means confined to the restoration of soil fertility and the development of the countryside. The dispossessed craftsmen started all kinds of industries, in which they used as labour-saving devices first water power, then the steam engine, the internal combustion engine, and finally electrical energy. By these agencies the Industrial Revolution, which continues till this day, was set in motion. It has influenced farming in many directions. In the first place, industries have encroached on and seriously reduced the area under cultivation. But by far the most important demand of the Industrial Revolution was the creation of two new hungers—the hunger of a rapidly increasing urban population and the hunger of its machines. Both needed the things raised on the land: both have seriously depleted the reserves of fertility in our soils. Neither of these hungers has been accompanied by the return of the respective wastes to the land. Instead, vast sums of money were spent in completely side-tracking these wastes and preventing their return to the land which so sadly needed them. Much ingenuity was devoted to developing an effective method of removing the human wastes to the rivers and seas. These finally took the shape of our present-day water-borne sewage system. The contents of the dustbins of house and factory first found their way into huge dumps and then into incinerators or into refuse tips sealed by a thin covering of cinders or soil.

At first the additional demands for food and raw materials were met by the restored agriculture and the periodical ploughing up of grass. One of these demands was the vast quantities of corn needed to feed the urban population. The price of wheat was regulated for more than 150 years by a series of Corn Laws, which attempted to hold the balance between the claims of the farmers who produced the grain and those of the consumers and the industrialists who advocated cheap food for their workers, so that they could export their produce at a profit. But as the urban population expanded, the pressure on the fertility of the soil increased until, in 1845, a disastrous harvest and the potato famine compelled the Government in 1846 to yield. The "rain rained away" the Corn Laws (Prothero).

Deprived of protection, farmers were forced to adopt new methods and to farm intensively. Many developments in farming occurred. Particular attention was paid to drainage: the first drain pipe was made in 1843; two years later the pipes were turned out by a machine. Liebig's famous essay in 1843 drew attention to the importance of fertilizers. While better farm buildings and the preparation of better farmyard manure were adopted, two fatal mistakes were made. Artificial fertilizers

54

like nitrate of soda and superphosphate came into use: imported feeding stuffs for livestock began to take the place of home-grown food. British farming, in adopting these two expedients, because they appeared for the moment to be profitable, laid the foundations of much future trouble. But in the use of better implements for the land and the provision of improved transport facilities the countryside was on firmer ground. The result of all these and other developments was a period of great prosperity for farming which lasted till late in the seventies of the last century.

THE GREAT DEPRESSION OF 1879

Then the blow fell. The year 1879, which I remember so vividly, was one of the wettest and coldest on record. The average yield of wheat fell to about fifteen bushels to the acre: large numbers of sheep and cattle were destroyed by disease: the price of wheat fell to an undreamt-of level as the result of large importations from the virgin lands of the New World. The great depression of 1879 not only ruined many farmers, but it dealt the industry a mortal blow. Farmers were compelled to meet a new set of conditions—impossible from the point of view of the maintenance of soil fertility—which have been more or less the rule till the Great War of 1914–18 and the World War which began in 1939 provided a temporary alleviation as far as the sale of produce and satisfactory prices were concerned.

Since 1879 the standard of real farming in Great Britain has steadily fallen. The labour force, particularly the supply of men with experience of and sympathy with livestock, markedly diminished and deteriorated in quality. Rural housing left much to be desired. Drainage was sadly neglected. The small hill farms, which are essential for producing cattle possessing real bone and stamina, fell on evil days. Our flocks of folded sheep, so essential for the upkeep of downland, dwindled. Diseases like foot-and-mouth, tuberculosis, mastitis, and contagious abortion became rampant. Less and less attention was paid to the care of the manure heap and to the maintenance of the humus content of the soil. The NPK mentality (p. 72) replaced the muck mentality of our fathers and grandfathers. Murdered bread, deprived of the essential germ, replaced the real bread of the last century and seriously lowered the efficiency of our rural population. The general well-being of our flocks and herds fell far below that of some of our overseas competitors like the Argentine.

But in this dark picture some rays of light could be detected. The pioneers were busy demonstrating important advances. Among these two are outstanding: (1) the Clifton Park system of farming based on deep-rooting plants in the grass carpet, and (2) the use of the subsoiler for breaking up pans under arable and grass, and so preparing the ground for another great advance—the mechanized organic farming of tomorrow.

THE SECOND WORLD WAR

Such, generally speaking, was the condition of British agriculture in September 1939, when the second world war began and the submarine menace for the second time brought national starvation into the picture. What an opportunity was provided for a Coke of Norfolk for making use of a portion of the resources of a great nation to set British farming on its feet for all time by the simple expedient of restoring and maintaining soil fertility! What an opening was given to the pioneers of human nutrition and the apostles of preventive medicine for feeding the men and women defending the country on the fresh produce of fertile soil and so initiating the greatest food reform in our history! But the potential Cokes of Norfolk had been liquidated or discouraged by many years of death duties, which had destroyed most of our agricultural capital and deprived the countryside of its natural leaders who, in years gone by, had done so much for farming. The apostles of real nutrition and of preventive medicine, such as the panel doctors of Cheshire, were ignored.

A much easier road was taken. The vast stores of fertility, which had accumulated after the long rest under grass, were cashed in and converted into corn crops. The seed so obtained saved the population from starvation, but most of the resulting straw could not be used because of the shortage of labour to handle it and of insufficient cattle to convert it into humus. The grow-more-food policy was, therefore, based on the exhaustion of the soil's capital. It is a perfect example of unbalanced farming. It is therefore certain to sow the seeds of future trouble, which will be duly registered by Mother Earth in the form of malnutrition and disease of crops, livestock, and mankind.

56

5

INDUSTRIALISM AND THE PROFIT MOTIVE

ONE OF THE developments which marks off the modern world is the growth of population. The figures are startling. There were about nine hundred million persons living during the eighteenth century, but over two thousand million at the beginning of the twentieth; in a century and a half world population, therefore, more than doubled. The principal increases took place in Europe.

The first effect of this is obvious—there were many more mouths to feed. Had no other changes accompanied this rise in population, we can guess what might have happened. The density of the peoples in rural Europe might have rivalled that in peasant China, and European agriculture would either have had to evolve methods of intensive cultivation similar to those of the Chinese or the additional population could not have survived.

Fate or their own ingenuity has sent the Western nations along another path. The picture has become quite different from that of the Far East and a very remarkable picture it is. We are so accustomed to it that we scarcely grasp the anomalies which it represents or the dangers into which it is leading us.

THE EXPLOITATION OF VIRGIN SOIL

The new populations did not, as a matter of fact, remain in Europe in their entirety. The Western peoples reached forth and put themselves in possession of vast areas of virgin soil in North America, Australia, New Zealand, and South Africa. Naturally agriculture became extensive, which word means that the cultivator prefers to get a smaller volume of produce per acre off a larger area rather than a great deal from a smaller area more intensively worked. The tracts seized were so enormous that each settler had at his disposal not a tiny piece of

57

ground from which to raise as much produce as possible, but a huge section—running into hundreds of acres for the growing of crops, into thousands for the raising of cattle or sheep. The amount of human effort to be put into each acre became indeed the crucial question—in contrast with Europe the new populations were thin and a thin population means few hands, and few hands can do little manual work. The first significant fact we have to note is the uneven distribution of the enlarged population as between the old and the new countries.

It was in these circumstances that the machine came to the help of agriculture. The outcome of the use of machines in farming was revolutionary; this is not always realized. Five men working with the most modern combine [1] can harvest and thresh fifty acres of wheat in the same number of hours as would require 320 persons working with old-fashioned hand tools; two men working with a header can replace 200 working with sickles; other calculations show for certain specified jobs only one-twentieth or even only one-eightieth of the amount of human labour formerly employed.[2] If these particular calculations apply exclusively to the easier processes of crop cultivation and reaping, it may also be pointed out that the cream separator and machine milking have effected a dramatic augmentation of the dairy industry by saving human labour.

We have reason to be grateful to those who invented the powerful devices which made possible these results. The food which has fed the great populations of Western civilization has been, in part, machine-produced food; without these machines such populations must have starved. But there is another side to the picture. The ease with which agriculture was mechanized was in itself a temptation and this temptation the Western nations have not been able to withstand. It has seemed so easy to provide enough food with comparatively little human labour, and not only this, but also to supply with raw materials those other machines, industrial in character and situated in manufacturing districts, which have been the invention of an ingenuity even more refined than has gone to the making of the agricultural harvester or combine. From these machines, continuously fed with the wool, cotton, silk, jute, hemp, sisal, rubber, timber, and the oil seeds of the whole world, has flowed a vast stream of industrial articles which have been at the dis-

---

[1] So called because it is a machine combining cutting and threshing. A header is another form of the combine.

[2] Howard, Louise E., *Labour in Agriculture* (Oxford University Press and Royal Institute of International Affairs, 1935), pp. 244–5.

posal of all and which have given a quite special character to our modern civilization.

The result has been inevitable. The hunger of the urban populations and the hunger of the machines has become inordinate. The land has been sadly overworked to satisfy all these demands which steadily increase as the years pass.

Not even the power of the machine would have been sufficient to feed and supply the immense populations of the nineteenth century had it not been for the vast natural capital in the shape of the humus stored in the soils of the new continents now opened up. The general exploitation of these soils did not take place until the nineteenth century was well on its way. Then the settlers who had poured westwards in North America, trekked northwards from the coast of South Africa, landed by the boatload in the harbours of New Zealand and Australia, set themselves to exploit this natural wealth with zest: they were eager to follow the covered wagon and to draw the plough over the prairies where once only herds of bison had roamed. Meanwhile in South and Central America, Ceylon, Assam, South India, the Dutch East Indies, and East Africa the plantation system, already known in the eighteenth century in the West Indies, took on a magnitude and an aspect which made it a new phenomenon. From all these sources immense volumes of food and raw materials reached Europe in such abundance that no one stopped to ask whether the stream could continue for ever.

Yet all these processes were almost pure harvesting, a mere interception and conversion of Nature's reserves into another form. It is true the land was tilled after a fashion, cultivated and sown, though in such industries as timber and rubber not even that, the ancient riches of the forest being for many years merely plundered. But whatever cultivation processes were undertaken did not amount to much more than a slight, necessary disturbance of those rich stores of accumulated humus which Nature had for hundreds of years been collecting under the prairie or the forest. So enormous were these reserves that the land bore crop after crop without faltering. In such regions as the great wheat belt of North America fifty years of wealth was available and the farmer knew well how to dig into these riches.

The phrase *mining the land* is now recognized as a very accurate description of what takes place when the human race flings itself on an area of stored fertility and uses it up without thought of the future. In the mid-nineteenth century this began to take place on an unprecedented

59

scale. For if agriculture was, so to say, the nurse of industry, she was persuaded to learn one salient lesson from her nursling. This was the lesson of the profit motive.

THE PROFIT MOTIVE

Of course, ever since the decay and final collapse of the feudal system, when service steadily gave place to rents, European agriculture has been working for profit; it was already in Tudor times a feature of the British wool trade which preceded and followed enclosure; the great English agricultural pioneers of the eighteenth century were also perfectly alive to the question of the monetary return for their reforms. Indeed, as soon as any harvest is sold rather than consumed, the question of profit must arise. The problem is one of degree and emphasis. Is profit to be the master? Is it to direct and tyrannize over the aims of the farmer? Is it to distort those aims and make them injure the farmers' way of living? Is it to be pushed even further and to make him forgetful of the conditions laid down for the cultivation of the earth's surface, so that he actually comes to defy those great natural laws which are the very foundation and origin of all that he attempts? If this is so, then the profit principle has outrun its usefulness: it has been dragged from its allotted niche in the world's economy, set on a high altar, and worshipped as a golden calf.

At first sight the profit motive does not seem to have taken modern farming very far. The farmers of the new countries opened up in the nineteenth century did not make vast fortunes. Perhaps in sheep farming and without doubt in the plantation industries large money was at one time made. But on the whole the monetary rewards of the new farming were not impressive. They never bore comparison with the colossal fortunes which nineteenth-century manufacture produced for the factory owner. Unlike the cotton spinner, the North American farmer did not exchange his shack for a huge and luxurious mansion. He remains to this day a dirt farmer, and is proud to call himself so, in close contact with his work and doing it with his own hands. It is, therefore, not easy to grasp that without great personal wealth and with no harmful intentions he was, nevertheless, a true despoiler, and that in so far as the occupation on which he was engaged is the first occupation in the world, while the means which he handled—the soil—is the most sacred of all trusts, he did more harm in his two or three generations than might be thought possible.

The ease with which crops could be grown year after year on new soil tempted the farmer to forget the law about restoring that fertility which he was rapidly using up in his farming operations. The soil responded again and again. Crop after crop of wheat was raised. Labour, as we have seen, was scarce and animals require much knowledge and much attention. As manure did not seem to be required, animals were discarded. Thus the straw could not be rotted down and the normal practice was to burn it off where it stood. In effect this was to repeat that old wasteful practice of the primitive shifting cultivator who renders the tropical forest into ash: in both cases a potentially rich organic matter was reduced to the inert inorganic phase and so deprived of its duty to the soil population. In short, the old mixed husbandry, which had maintained Europe and which not long before the settlers migrated had been so notably improved as really to achieve something approaching a balance of the processes of growth and decay, was never brought across the waters—its principles slipped from the settler's mind: he was unaware of his loss.

THE CONSEQUENCE OF SOIL EXPLOITATION

The result of the exploitation of the soil has been the destruction of soil fertility on a colossal scale. This has taken place in the areas to which we have been referring at different rates over different periods and in response to various factors. The net result of a century's mismanagement in the United States was summed up in 1937 as either the complete or partial destruction of the fertility of over 250,000,000 acres, i.e. 61 per cent of the total area under crops: three-fifths of the original agricultural capital of this great country has been forfeited in less than a century. But New Zealand where a systematic burning of the rich forest to form pasture which in its turn was soon exhausted, parts of Africa where overstocking has ruined much natural grazing, Ceylon where a criminal failure to follow the native practice of terracing for rice has denuded the mountain slopes of their glorious forest humus, would probably show consequences just as startling. Almost everywhere the same dismal story could be related.

When stockbreeding in its turn began to offer strong monetary inducements, especially in Australia and New Zealand in the 1880's and 1890's, another phase set in. Animals were kept in enormous numbers —some sheep runs owned hundreds of thousands of sheep—but scant regard was paid to their nurture; the natural herbage, untouched for centuries, was counted upon and as long as the humus held out such specialized animal husbandry could continue. But when the stores of humus were worked out, trouble began. Disease appeared. Inevitable accidents, especially drought, brought utter disaster: there was colossal mortality. No doubt Nature is prepared for such waste: but man is not. It is a setback for him. The right provision against such emergencies would have been a reserve of fodder in the form of cultivated roots or hay, for drought kills not so much by want of water as by starvation. But as crops were not grown alongside of the animals, there were no such reserves; while the natural remedy of wandering to a new pasture, which might have mitigated the catastrophe for the much smaller numbers of wild animals, was no longer possible. Thousands of sheep or cattle therefore perished: the profit motive had become a boomerang.

As the years have passed, the toll of animal disease has become so severe that Governments feel obliged to compute it statistically and grasp at all remedies. The figures rival in their intrinsic importance the figures of erosion. Actually it is the same bad effect in each case: we are looking at the results of mono-crop farming so called.

Let us recall our examination of the methods of Nature. We had noted among other things that her mechanisms for dispelling and scattering seeds were singularly perfect. Is it not obvious that Nature refuses to grow on any one spot the same crop without other intermixtures? Some aggregation of identical plants may take place: so does some collection of animal life: Nature knows the herd, the swarm—these are her own inventions, but they are set to carry out their lives in a mixed environment of other existences. It is to be noted that in the case of animals their natural range is great, involving change of habitat. It is also, perhaps, worth pondering over that when Nature does breed in one locality a large number of the same animals, these aggregations are particularly liable to be decimated by such diseases as she chooses to introduce; it is as though she herself repented of this principle of aggregation and in her own ruthless way chose for the time being to terminate it. But allowing for these slight modifications, the general economy of Nature is mixed in an extraordinary way. Her sowings and harvestings are intermingled to the last degree, not only spatially, but in succession of time, each plant seizing its indicated opportunity to catch at the nu-trient elements in air, earth, or water, and then giving place to another, while some phases of all these growing things and of the animals, birds, and parasites which feed on them are going on together all the time. Thus the prairie, the forest, the moor, the marsh, the river, the lake, the ocean include in their several ways an interweaving of existences which is a dramatic lesson; in their lives, as in their decay and death, beasts and plants are absolutely interlocked. Above all, never does Nature separate the animal and vegetable worlds. This is a mistake she cannot endure, and of all the errors which modern agriculture has committed this aban-donment of mixed husbandry has been the most fatal.

It would be to distort the picture unfairly if we were to assume that these mistakes were to be found only in the farming of the new countries. That was by no means the case. The thirst for profit profoundly affected European husbandry also. The yield became everything; quality was sacrificed for quantity. The merest glance at any recent set of agricul-tural statistics will reveal how wholly this factor of quantity is now in-sisted upon, indeed is made a boast. Rises in the yield of cereals per acre are everlastingly cited; yields of milk per cow become an obsession. There is, no doubt, virtue in increased volume of produce; it is the aim of agriculture to produce largely, and such increase is useful to man-kind. But if the profit and loss account is made to look brilliant merely because capital has been transferred and then regarded as dividend, what business is sound?

THE EASY TRANSFER OF FERTILITY

The using up of fertility is a transfer of past capital and of future possibilities to enrich a dishonest present: it is banditry pure and sim-ple. Moreover, it is a particularly mean form of banditry because it in-volves the robbing of future generations which are not here to defend themselves.

It is, perhaps, not realized over what distances the transfer of fertility can now take place. This final aspect is an unforeseen consequence of the vast improvement in means of communication. It is not necessary for the modern farmer to cash in his own fertility to make a good in-come; he has a more subtle means at hand. Before the present world war the telephone farmer, as he was sometimes called, had merely to ring up his agent and the needed quantity of imported foodstuffs, oil-cakes, or whatever it may be, was delivered by truck the next morning. It was claimed that the dung of his animals was thereby enriched and that whatever fields he condescended to cultivate were thus improved. This is true. But what does it amount to? Merely that the accumulated fertility of those distant regions of the earth which have produced the materials for the oil-cake is being robbed in order to bolster up a worn-out European soil: the same bad process of exhaustion is going on, but at the moment so far away that it can be temporarily ignored. On such a system of imported foodstuffs the whole of the dairy industry of Denmark was built up. The Danish farmer was not carrying on agriculture at all: he was devoting himself to a mere finishing process and what he built up was a conversion industry. It is an astonishing sidelight that before the present war the Danish farmer frequently sold his good butter to the London market and bought the cheaper margarine for his children's use. The pursuit of profit had invaded not only his farming methods but his way of life and had even encroached on the health and well-being of his family.

The transfer of fertility to current account, as it were, has not ceased: soil erosion and the toll of animal disease continue. Two recent writers calculate that erosion is even now proceeding "at a rate and on a scale unparalleled in history": between 1914 and 1934, they declare, more soil was lost to the world than in all the previous ages of mankind,[1] while a host of learned papers are evidence that new diseases of stock are being discovered day after day, baffling both farmer and veterinary surgeon.

The remedy is simple. We must look at our present civilization as a whole and realize once and for all the great principle that the activities of *homo sapiens,* which have created the machine age in which we are now living, are based on a very insecure basis—the surplus food made available by the plunder of the stores of soil fertility which are not ours but the property of generations yet to come. In a thoughtful article by Mr. H. R. Broadbent recently published in the *Contemporary Review* (December 1943, pp. 361–4) this aspect of progress is discussed and the conclusion is reached that:

"The whole world has shared, either directly or indirectly, with the United States and British Commonwealth of Nations in the use of the surplus from the eroded lands. It has enabled us to build up our engineering knowledge and technique. Our buildings, engines, and machin-

[1] Jacks, G. V.,and Whyte, R. O., *The Rape of the Earth* (Faber and Faber, London, 1939).

ery are material evidence of its consumption; but the foundation has been impoverishment of the soil. The food was cheap—the products were cheap because the fertility of the land was neglected. We in England have often been puzzled by the arrival of cheap goods when it was known that high wages were paid to the makers. We had not seen the land which had produced not only the food for those makers, but also the organic material which they processed. . . . We had not seen the gullies torn out from the land by unabsorbed rains and melting snows. We had not seen the dust storms of the wind seeping out the goodness from the soils and carrying it hundreds of miles from its old resting place. When we look on Battersea Power Station or our reclaimed land, the great railroads of the United States or London's Underground, or consider such wonders as the general use of electricity and mechanical transport, the spread of broadcasting and mass-production of clothes, we must also see the devastated lands which have yielded the surplus to make them possible. These things in which we take pride were built on an unbalanced surplus, the unmaintained capital of the soil. No country can continue indefinitely to provide food and material at such a cost. Under extraordinary conditions, as in war, the land must be driven beyond the normal to provide an extravagant surplus. But war is abnormal, and the normality at which we aim is peace which implies stability of foundations. Raymond Gram Swing broadcast that at the rate of soil and water depletion occurring when the 1934 survey was made in fifty years the fertile soil of the United States would be one-quarter of what was present originally, and that in a hundred years at the same rate of depletion the American continent would turn into another Sahara. Perhaps he was thinking of other civilizations buried in the sands; the ruins of ancient towns and villages in the Gobi desert, Palestine, and Mesopotamia. Perhaps he feared the fate of the country north of the Nigerian boundary, where an area as large as the Union of South Africa has become depopulated in the last two hundred years. Perhaps he remembered the malaria-ridden marshes of Greece and Rome which came with the decline of their agricultural population and loss of vigour."

THE ROAD FARMING HAS TRAVELLED

What is the outcome of our arguments? We started our investigations by considering the operations of Nature and continued them by summarizing human action in relation to those operations. It is our actions, when confronted with forms of natural wealth, which have shaped the modern world in its economic, financial, and political contours. The harvesting, distribution, and use of natural resources is the first condition which determines human societies.

The supplies provided by Nature are the starting point for everything. Primitive societies have to adapt themselves to what supplies lie readily to hand; they sometimes use severe processes of self-correction, e.g. infanticide, in order to do so. But a further stage is usually reached. Nature's supplies are not static; they appear as actual surpluses, and by a bold use of these surpluses societies emerge from the primitive stage. This use later becomes crystallized as the profit motive.

To eliminate this would be impossible. In advanced societies it would be a retrograde step. The profit motive, however far it may have led us astray, is founded on physical realities. It is wiser to go back to those realities, reconsider them, and seek any necessary correction from a better understanding of them.

What are the exact conditions attaching to the creation of the surpluses which Nature accumulates?

In spite of the fact that we speak of her lavishness, Nature is not really luxurious: she works on very small margins. Natural surpluses are made up of minute individual items: the amount contributed by each plant or animal is quite tiny: it is the additive total which impresses us. The further result is that the gross amounts of these surpluses are not disproportionate to their environment: harvests are only a small part of natural existences.

The farmer is apt to disregard these facts. His object is to produce more. It pays him to select a smaller number of plants or animals and make each of these produce more intensively: he counts on the elasticity of Nature. If he kept his harvests to the very small proportions usual in wild existences, his farming would be exceedingly laborious and scarcely worth while: farming improves in proportion to the extra amounts which the cultivator manages to elicit by stimulating rates and intensities of growth. Up to a point he can do this with safety. After that Nature refuses to help him: she simply kills off the over-stimulated existence. Her elasticity is great, but it is not infinite.

Here we may find our principal warning. The pursuit of quantity at all costs is dangerous in farming. Quantity should be aimed at only in strict conformity with natural law, especially must the law of the return

66

of all wastes to the land be faithfully observed. In other words, a firm line needs to be drawn between a legitimate use of natural abundance and exploitation.

Modern opinion is now set against all forms of exploitation. The limitation of money dividends, the disciplining of capital investments have begun. Undertaken originally only from the point of view of economic order, then continued for political and national motives, these measures bear in themselves further possibilities; it would be easy to give them wide moral significance.

In agriculture, which is so much more fundamental than industrial economics, the field is still uncharted. The agricultural expert still holds out the ideal of quantity as the highest aim. Helpless under this leadership, the farmer has first himself been exploited and has then almost automatically become an exploiter. A vicious round has been set up, resistance to which is only just showing itself.

The first pressure has been the pressure of urban demand. This pressure is of long standing and has been very greedy. It has been exercised in strange contradiction to another tendency: while the farmer was asked to produce more, the man-power needed for greater production was enticed away to the cities, there to add to the number of mouths to be fed. The farmer was always being asked to do more with less man-power to do it. This absurdity has not passed unnoticed. Severe criticisms have been enunciated; everyone would agree to any reasonable measures to restore the balance of population. That the balance of physical resources has also been disturbed is only just beginning to be realized. The transference of the wealth of the soil to the towns in the shape of immense supplies of food and raw materials has not been made good by a return of town wastes to the country. This return is a *sine qua non* and should at all costs include the crude sewage, which is by no means impossible even with modern systems of drainage. If this can be arranged, the existence of cities will cease to be a menace: exploitation will stop, legitimate use will return. Nevertheless, it will always be important to exercise some control over the volume of urban demand, probably by some restrictions on the size of the urban community, which means some restrictions on the launching of new industries or the expansion of old ones. However far off this sort of control may seem at the present time, it must at some future date rank among the preoccupations of the statesman. Otherwise there will never be any protection for the farming world from the incredible demand for quantity.

67

It has been under the pressure of this insatiable demand that the farmer has himself become an exploiter: in two ways. Having exhausted the possibilities of production from his own fields, he has actually had the temerity to transfer to those fields the stored-up natural wealth, representing centuries of accumulation, lying many thousand miles away. The importation of feeding stuffs, of guanos and manures of all kinds from distant parts of the world to intensify European farming is only robbery on a vast scale. It is not necessary to claim that every national agriculture must be completely self-contained: this would be a great pity. But the tide has been all one way. While from the economic and financial point of view the return flow of manufactured goods is supposed to be a *quid pro quo,* from the point of view of ultimate realities this type of return is perfectly useless. The draining away of natural fertility from tropical and sub-tropical regions is exceedingly dangerous. It is a point on which the peoples of these regions may later come to put a colossal question to the conscience of the so-called civilized countries: Why has the stored-up wealth of our lands been taken away to distant parts of the world which offer us no means of replacing it?

Even this dangerous expedient has been insufficient. Faced with the demand for higher yields, the farmer has grasped at the most desperate of all methods: he has robbed the future. He has provided the huge output demanded of him, but only at the cost of cashing in the future fertility of the land he cultivates. In this he has been the rather unwilling, but also the rather blind, pupil of an authority he has been taught to respect: the pundits of science have urged him to go forward and have made it a matter of boasting that they have done so. How this has come about will be described in our next chapter.

6

THE INTRUSION OF SCIENCE

I⊤ was Francis Bacon who first observed that any species of plants impoverished the soil of the particular elements which they needed, but not necessarily of those required by other species. This true observation might have put subsequent investigators on the right path had their general knowledge of scientific law been less fragmentary. As it was, many ingenious guesses were made in the course of the seventeenth and eighteenth centuries as to the nurture and growth of plants, some near the truth, some wide of the mark. Confusedly it began to be recognized that plants draw their food from several sources and that water, earth, air, and sunlight all contribute. Priestley's discovery of oxygen towards the end of the eighteenth century opened up a new vista and the principles of plant assimilation soon came to be firmly established, by which is meant the fact that under the influence of light the green leaves absorb carbon-dioxide, break it up, retaining the carbon and emitting the oxygen (hence their purifying effect on the atmosphere)—what is more delicious than the air of the forest, garden, or field?—while without light, i.e. during the night-time, plants reverse the process and emit carbon-dioxide. Though the investigation of the parallel processes of root respiration, i.e. the use made by the roots of the oxygen available from the soil-air or the soil-solution, did not follow until a good deal later, yet the foundations of knowledge about the life of plants were at least thus laid on sound lines.

THE ORIGIN OF ARTIFICIAL FERTILIZERS

It was at this juncture that a special direction was given to investigation by Liebig. Liebig is counted the pioneer of agricultural chemistry. His *Chemistry in Its Application to Agriculture,* contributed to the British Association in 1840, was the starting point of this new science. His inquiries into general organic chemistry were so vast and so illuminating that scientists and farmers alike naturally yielded to the influence of his teaching. His views throughout his life remained those of a chemist and he vigorously combated the so-called humus theory, which attributed the nourishment of plants to the presence of humus. At that time the soil in general and the humus in it were looked on as mere collections of material without organic growth of their own; there was no conception of their living nature and no knowledge whatever of fungous or bacterial organisms, of which humus is the habitat. Liebig had no difficulty in disproving the role of humus when presented in this faulty way as dead matter almost insoluble in water. He substituted for it a correct appreciation of the chemical and mineral contents of the soil and of the part these constituents play in plant nourishment.

This was a great advance, but it was not noticed at the time that only a fraction of the facts had been dealt with. To a certain extent this narrowness was corrected when Darwin in 1882 published *The Formation of Vegetable Mould Through the Action of Worms, with Observations of Their Habits,* a book founded on prolonged and acute observation of natural life. The effect of this study was to draw attention to the extraordinary cumulative result of a physical turnover of soil particles by natural agents, particularly earthworms. It was a salutary return to the observation of the life of the soil and has the supreme merit of grasping the gearing together of the soil itself and of the creatures who inhabit it. Darwin's book, based as it is on a sort of experimental nature study, established once for all this principle of interlocked life and, from this point of view, remains a landmark in the investigation of the soil.

Meanwhile Pasteur had started the world along the path of appreciating the marvellous existence of the microbial populations traceable throughout the life of the universe, unseen by our eyes but discoverable to the microscope. The effect of his investigations has been immense; enormous new fields of science have been opened up. The application of this knowledge to agriculture was only gradual. Many years slipped by before it was realized that the plants and animals, whose life histories are based ultimately on living protoplasm, have their counterparts in vast families and groups of miscroscopic flora and fauna in the very earth on which we tread.

It thus came about that the chemical aspects of the soil for a long time predominated in the mind of the scientist. The theory had had a good start, it was older and naturally better developed. Moreover, and this is important, Liebig had been a pioneer not only in science, but in practice. From the outset of his experiments he had made every effort to work with the farmer and also by field investigation. The farmer did not object to the help given him in his difficult task. As the demands on him grew to fever pitch, for he was just facing the heavy, cumulative greed of the expanding factories of the world and the hunger of their servants, the workers, he not unnaturally welcomed ideas and suggestions which he was told would enable him to carry out his task in an easy, practical, and clean way without fuss and without that extra labour already so difficult to procure.

Thus artificial fertilizers were born out of the abuse of Liebig's discoveries of the chemical properties of the soil and out of the imperative demands made on the farmer by the invention of machinery. It must be confessed that Liebig himself was somewhat of a sinner on this count. He manufactured artificial manures and though these were oddly enough a failure he maintained his faith, which indeed was questioned by none, that the food of plants could be replenished by the too obvious principle of putting back into the earth the minerals which, as the analysis of the ash of the burnt crops taken off it revealed, were drawn out by the plants.

As long as this principle was held to override every other consideration, no further progress could be made. The effects of the physical properties of the soil were by-passed: its physiological life ignored, even denied, the latter a most fatal error. There was a kind of superb arrogance in the idea that we had only to put the ashes of a few plants in a test tube, analyse them, and scatter back into the soil equivalent quantities of dead minerals. It is true that plants are the supreme, the only, agents capable of converting the inorganic materials of Nature into the organic; that is their great function, their justification, if we like to use that word. But it was expecting altogether too much of the vegetable kingdom that it should work only in this crude, brutal way; as we shall see, the apparent submission of Nature has turned out to be only a great refusal to have so childish a manipulation imposed upon her.

At first all seemed to go well. As economic conditions pressed on the farmer more and more severely, he thankfully grasped at the means of increasing the volume of his production and after the great agricultural depression of 1879 began to use the artificial manures placed on the market for his benefit. These were of two kinds; the nitrogen artificials which supply the current account of plants and which have a marked effect in increasing leafage, and the potash and phosphate artificials which increase the mineral reserves of the soil. The chemical symbol for nitrogen is N; for potassium, K (for Kalium); and for phosphorus, P; and the attitude of mind which sees all virtue in the use of artificials may fairly be dubbed the NPK mentality.

THE ADVENT OF THE LABORATORY HERMIT

Stimulating the growers who began to acquire this mentality, there came to be installed in the strongholds of science a type of investigator whom we are justified in naming the laboratory hermit. The divorce between theory and practice was a new phase which would have been deprecated by Liebig, but the temptation to grow a few isolated plants in pots filled with sand—watered by a solution containing the requisite amount of NPK in a balanced form so that any one constituent did not outdo the others—draw them, measure them, tie them up in muslins, weigh them, burn them, and analyse them proved too great. A quantity of minute investigation was based on these practices, which are only justified as a mere introduction to agricultural investigation. Though the plant may to some extent be grown under these conditions, the soil is another problem. Soil or watered sand in a flower-pot is literally in a strait-jacket and it is nonsense to assume that it can carry on its proper life: for one thing the invasion of earthworms or other live creatures is eliminated and many other processes put out of action. That essential co-partnership between the soil and the life of the creatures which inhabit it, to which Darwin's genius had early drawn attention, is wholly forgotten.

To confirm the findings of the flower-pots the small plot trials—in which some fraction of an acre of land is the usual unit—were devised. Great virtues have been attributed to the repetition of such tests over a long period of years and, of late, to the statistical examination of the yields. In this way it was hoped to "disentangle the effects of various factors and to state a number of probable relationships which can then be investigated in the laboratory by the ordinary single factor methods." [1]

THE UNSOUNDNESS OF ROTHAMSTED

At this point the manifold weakness of the small-plot method of agricultural investigations must be emphasized. The celebrated Broadbalk wheat trials at Rothamsted, the units of which are strips of land

[1] Russell, Sir John, *Soil Conditions and Plant Growth* (London, 1937), p. 31.

some half an acre in size and on whose results the artificial manure industry is largely founded, can be taken as an example. The trials have been repeated for some hundred years, the work has been carried out with extreme care, the fullest records have been kept and preserved, and the final figures have been subjected to the best available statistical analysis.

The main object of these experiments was to determine whether wheat could be grown continuously by means of artificials alone or with no fertilizer, and also to compare the results obtained by chemicals on the one hand and by farmyard manure on the other. The results are considered to prove that under Rothamsted conditions satisfactory yields of wheat can be obtained by means of chemicals only, that no outstanding advantage follows the use of farmyard manure, and further that on the no-fertilizer plot a small but constant yield of grain can be reaped. A subsidiary, but very important, result is also claimed, namely, that the fertilizing has had no appreciable effect on the quality of the wheat grain.

In spite of all the devotion that has been lavished on these Broadbalk trials, at least four major mistakes have been made in their design and conduct which completely discredit the final results.

In the first place, an error in sampling was made at the very beginning. A small plot cannot possibly represent the subject investigated, namely, the growing of wheat, which obviously can best be studied in this country on a mixed farm. We cannot farm a small strip of wheat land year after year, because it is difficult to cultivate it properly; the area does not come into the usual rotations and is, therefore, not influenced by such things as the temporary ley, by the droppings of livestock, and by periodic dressings of muck. The small plot, therefore, cannot represent any known system of British farming, any of our farms, or even the field in which it occurs. It only represents itself—a small pocket handkerchief of land in charge of a jailor intent on keeping it under strict lock and key for a century; in other words, it has fallen into the clutches of a Gestapo agent. In this sinister sense the Broadbalk trials have indeed been permanent.

In the second place, the continuous cultivation of wheat on a tiny strip of land is certain to create practical difficulties. Such land cannot be kept free from weeds because of the short time available between harvest in August and re-sowing in October. No cleaning crops like roots crop can, therefore, be used. This difficulty duly happened at

73

Rothamsted. The weeds got worse and worse and finally won the battle. Mother Earth rejected the idea underlying the continuous wheat experiment. The original conception of these trials has had to be modified. Fallows have had to be introduced. I last saw these Broadbalk plots about 1918 when this weed difficulty was causing considerable concern. I can truthfully say that never in my long experience have I seen arable land in such a hopeless and filthy condition. A more glaring example of bad farming could scarcely be imagined. I took my leave at the earliest possible moment and decided then and there that my last visit to Rothamsted—the Mecca of the orthodox—had been paid.

In the third place, no steps were taken to isolate the plots from the surrounding areas and to prevent incursions from burrowing animals such as earthworms. It is known from the work of Dreidax (*Archiv für Pflanzenbau*, 7, 1931, p. 461) and others on the Continent that when the earthworm population is destroyed by artificials, the affected areas are soon invaded by a fresh crop of worms from the neighbouring land. This invasion may take place at the rate of many yards a year. To study the effects of artificials on earthworms Dreidax showed that the experimental area should be at least ten acres and that the fringes of this land should never be taken into account. We know that artificials, sulphate of ammonia in particular, destroy the earthworm population wholesale; [1] but that after the nitrification of this manure has taken place the area is again invaded by more of these animals. A small oblong strip about half an acre in size is, therefore, obviously useless for determining the effect of artificials on the soil population. The unit should be a square at least ten acres in area. This wholesale destruction of the earthworm probably helps to explain the failures in wheat growing which often attend the application of the Rothamsted methods to large areas of land. The lowly earthworm—the great conditioner of the food materials for healthy crops—is murdered and no effective substitute is provided.

In the fourth place, the fertilizing scheme has never been allowed to impress itself on the variety of wheat grown. The fertilizing has influenced the soil, but not the plant. *The seed used every year has been obtained from the best outside source.* The wheat raised on each plot has not been used to sow that plot for the next crop. The plant has had a fresh start every sowing. The Broadbalk experiment is, therefore, not a

[1] The use of sulphate of ammonia for destroying earthworms on golf putting greens is recommended in *Farmers' Bulletin 1569* issued by the United States Department of Agriculture.

74

continuous wheat experiment as regards one of the two most important factors in the trial—the wheat plant itself. How this error crept in is difficult to say. It was most probably due to over-emphasis on the soil factor. Its discovery is largely due to Mr. H. R. Broadbent, who has made a critical study of the published reports on the Broadbalk plots from the beginning with a view to discovering the cause of the discrepancy between the Rothamsted experience and the results of large-scale wheat growing when carried out on the farm. In the discussions which arose Mr. Broadbent asked me where the seed sown every year on these plots came from. As this important fact was not recorded in the various Rothamsted Annual Reports, I asked the authorities to let me know the source of the seed used in the Broadbalk trials and was promptly informed that fresh seed was obtained every year from the best outside source and that the crop from each plot was never used to re-sow that plot. This candid confession invalidates the entire Broadbalk experiment. Had the harvest of each plot been used for re-sowing, in a very few years an important result would have been obtained. The effect of artificial manures, which we know is cumulative, would soon have begun to influence the stability of the variety itself and cause it to run out. In some period between twenty-five and fifty years the wheat would have ceased to grow and the Broadbalk experiment would have collapsed. This dramatic result, in all probability, would have saved the agriculture of this country and of the world from one of its greatest calamities—the introduction of artificial manures into general practice.

ARTIFICIALS DURING THE TWO WORLD WARS

In 1914, when the first world war broke out, the Broadbalk results were universally accepted as a safe guide for the farmer in the drive for increased food production. But it was the after-effects of this war rather than the four years of the war itself which ushered in a yet more ardent use of artificial fertilizers. The new process of fixing, i.e. combining, nitrogen from the air had been invented and had been extensively employed in the manufacture of explosives. When peace came, some use had to be found for the huge plants set up and it was obvious to turn them over to the manufacture of sulphate of ammonia for the land. This fertilizer soon began to flood the market.

From 1918 onwards the application of artificials was earnestly advocated by all authorities; their use was laid on the farmer almost as a moral duty. The universities had by now been impelled to set up agricultural departments, and finely equipped experiment stations were scattered over the various countries which in their general theory of investigation copied the universities, from which, indeed, they were invariably recruited. All these agencies without exception gave unconscious stress to the NPK mentality and were also hypnotized by the thraldom of fear of the parasite. Two thoroughly unsound and even mischievous principles thus acquired the support of the republics of learning—the universities—and the sanction of science itself. When the present war broke out the stage was set for the next swift advance towards the steep places leading downwards to the sea.

When, towards the end of 1939, the menace of the submarine began to imperil our food supplies from overseas, it became crystal clear that the fields of Great Britain would have to grow more and more of our nourishment if starvation were to be avoided. Then for the first and perhaps for the last time artificial manures came into their own: they were available in quantity to stimulate the crops: the Defence Regulations could be invoked to support the grow-more-food policy: the financial resources of a great nation were available to help the farmers to purchase these chemical stimulants and thus indirectly to subsidize the artificial manure industry itself: the staffs of these vested interests were at the disposal of the Ministry of Agriculture: the local War Agricultural Executive Committees soon became salesmen of the contents of the fertilizer bag: the frequent speeches of the Minister of Agriculture invariably contained some exhortation to use more fertilizers. The amalgamation of the vested interests and the official machine which directed war farming became complete. One thing, however, was forgotten. No satisfactory answer to the following question has been provided: What will be the final result of all this on the land itself, on the wellbeing of crops, livestock, and mankind? Will the grow-more-food policy have solved one problem—the prevention of starvation—by the creation of another—the enthronement of the Old Man of the Sea on farming itself? What sort of account will Mother Earth render for using up the last reserve of soil fertility and for neglecting her great law of return? Who is going to foot the bill?

THE SHORTCOMINGS OF PRESENT-DAY AGRICULTURAL RESEARCH

But the enthronement of the NPK mentality is only one of the blunders for which the experiment stations must be held responsible. The usual sub-division of science into chemical, physical, botanical, and other departments, necessary for the sake of clarity and convenience in teaching, soon began to dominate the outlook and work of these institutions. The problems of agriculture—a vast biological complex—began to be subdivided much in the same way as the teaching of science. Here it was not justified, for the subject dealt with could never be divided, it being beyond the capacity of the plant or animal to sustain its life processes in separate phases: it eats, drinks, breathes, sleeps, digests, moves, sickens, suffers or recovers, and reacts to all its surroundings, friends, and enemies in the course of twenty-four hours, nor can any of its operations be carried on apart from all the others: in fact, agriculture deals with organized entities, and agricultural research is bound to recognize this truth as the starting point of its investigations.

In not doing this, but adopting the artificial divisions of science as at present established, conventional research on a subject like agriculture was bound to involve itself and magnificently has it got itself bogged. An immense amount of work is being done, each tiny portion in a separate compartment; a whole army of investigators has been recruited, a regular profession has been invented. The absurdity of team work has been devised as a remedy for the fragmentation which need never have occurred. This is nonsensical. Agricultural investigation is so difficult that it will always demand a very special combination of qualities which from the nature of the case is rare. A real investigator for such a subject can never be created by the mere accumulation of the second rate.

Nevertheless, the administration claims that agricultural research is now organized, having substituted that dreary precept for the soul-shaking principle of that essential freedom needed by the seeker after truth. The natural universe, which is one, has been halved, quartered, fractionized, and woe betide the investigator who looks at any segment other than his own! Departmentalism is recognized in its worst and last form when councils and super-committees are established—these are the latest excrescences—whose purpose is to prevent so-called overlapping, strictly to hold each man to his allotted narrow path and above all to enable the bureaucrat to dodge his responsibilities. Real organization

77

always involves real responsibility: the official organization of research tries to retain power and avoid responsibility by sheltering behind groups of experts. The result of all this is that a mass of periodicals and learned papers stream forth, of which only a very few contain some small, real contribution.

The final phase has been reached with the letting loose of the fiend of statistics to torment the unhappy investigator. In an evil moment were invented the replicated and randomized plots, by means of which the statisticians can be furnished with all the data needed for their esoteric and fastidious ministrations. The very phrase—statistics and the statistician—should have been a warning. It is, of course, true and known to most persons that average numbers and similar calculations are not perfect; they are subject to various errors. Care is needed in interpreting them and, above all, experience of the actual: where this is available and where common sense is the judge, danger ceases. The deduction would be, in what we are now reviewing, that the agricultural investigator must be well acquainted with practical farming and be prepared to put his conclusions to practical tests over some period of time before he can be certain of what he says. This conclusion is just, and with such a corrective, agricultural experiment can live and prosper.

But the exactly opposite conclusion has been drawn. Instead of sending the experimenter into the fields and meadows to question the farmer and the land worker so as to understand how important quality is, and above all to take up a piece of land himself, the new authoritarian doctrine demands that he shut himself up in a study with a treatise on mathematics and correct his first results statistically. The matter has been pursued with zest and carried to all extremes; it is popularly rumoured that only one highly qualified individual is now able to interpret the mathematical principles on which are based the abstruse mass of calculations to which even the simplest experiments give rise.

But the proof of the pudding is in the eating thereof. Can the statistician give any practical help when the use of small plots gets into difficulties? In one case I personally investigated about 1936 the answer is: Most emphatically, no. This occurred at the Woburn Experiment Station, a branch of Rothamsted. During the summer I was invited by the Vice-President of the Rothamsted Trust, the late Professor H. E. Armstrong, F.R.S., to help him to discover why one of the sets of permanent manurial experiments at Woburn had come to an end. After a long treatment with artificials the soils on the greensand had gone on strike:

the cereals refused to grow. Why? I have a vivid recollection of this visit. We were first given a learned lecture on the past history of the plots with tables and curves galore by the Officer-in-Charge. We then visited the field, for which the professor said I was certain to need a spade. We saw the plots which had given up the struggle. No crop was to be seen, only a copious growth of the common mare's tail (*Equisetum arvense*). I then inquired whether a really good crop could be seen on similar land. We were shown a fine crop of lucerne nearby which had been manured with copious dressings of pig muck. The cause of the going on strike of the Woburn plots was now clear and the cure was obvious, but before explaining all this to the Officer-in-Charge I inquired what had been done by the Rothamsted staff to elucidate this trouble. It appeared that all the data and all the information available had been laid before the Director and his staff, including the statisticians, but without result. Neither the official hierarchy nor the higher mathematics had any explanation or advice to offer. I thereupon explained the cause and pointed to the cure of the mischief. Constant applications of chemicals to this sandy soil had so stimulated the soil organisms that the humus, including the humic cement of the compound soil particles, had been used up. This had led to pan formation and to the cutting off of the air supply to the subsoil. All this was obvious by the establishment of a weed flora mostly made up of a species of *Equisetum*. My diagnosis would be confirmed by an examination of the soil profile which would disclose a sand pan some six to nine inches below the surface and the development of the characteristic root system of this weed of poorly aerated soils. This injurious soil condition could be removed by a good dressing of muck followed by a crop of lucerne. A soil profile was then exposed and there was the pan and the root system exactly as I foreshadowed. It was merely a case of reading one's practice in the plant. The establishment of the mare's tail on a high-lying sand could only be possible by poor soil aeration due, in all probability, to the formation of a subsoil pan so common in sandy soils. Farmyard manure, plus a deep-rooting crop and earthworms, would prevent pan formation, hence the good crop of lucerne. Long practical experience and many years spent in root studies had instantly suggested the cause of the Woburn trouble. Many years' observation and first-hand experience of the lucerne crop enabled me to suggest a cure for the pan formation. How could statistics and the higher mathematics be a substitute for the faculty of reading one's practice in the plant?

79

How could this faculty be developed except by a wide experience of research methods and of practical farming?

Can statistics or the statistician help in unravelling the nature of quality—that factor which matters most in crop production, in animal husbandry, and in human nutrition? We cannot measure or weigh quality and express the result in numbers which the statistician can use. But our livestock instantly appreciate quality and show by their preference, their better health, their improved condition and breeding performance how important it is. The animal, therefore, is a better judge of one of the factors that matters most in farming than the mathematician. But on this important point—the verdict of the animal—the records of our experiment stations are silent. At these institutions crops are weighed on metal or wooden balances so that figures—the food of the statistician—can be provided. But if many of these experimental crops, particularly those raised with chemical manures, are tested in the stomachs of our livestock—the real balance of the farmer—they will be found wanting.

The invasion of statistics into agricultural research has been an incursion into a diseased field. Let us sum up this chapter by judging this result of our modern civilization by its works. This surely is not unfair. Of some fifteen committees set up in Great Britain under the Agricultural Research Council just before the recent war no less than twelve were allocated to investigation of the diseases of animals and plants. Of the enormous mass of scientific literature published on agricultural problems some third part is concerned with the onset, history, description, or attempted remedies for some form of sickness or disability in crops or livestock. This merely reflects the facts. Old diseases are spreading and new diseases are appearing. Eelworm devours our potato crop, foot-and-mouth disease infects our cattle, grass sickness kills our horses, fungi, viruses, and insect parasites invade our fruit and our vegetables: every vine in France is smothered in green and blue copper compounds to keep the mildews at bay. Comparatively new crops like the sugar beet are now retreating before the onset of the eelworm. New scientific organizations and their satellite companies for dealing with the increasing manufacture and sale of insecticides and fungicides are being created. The farmers are being urged to subscribe to panels of veterinarians to control the growing toll of disease among their livestock.

Even a Beveridge health plan is now being advocated by the National Veterinary Association, who also favour "the establishment of State breeding farms to facilitate the improvement of average stock by direct mating and by controlled artificial insemination" (*Daily Express*, 16th March 1944). The practice of artificial insemination for livestock can only be described as a monstrous innovation which can only end in life-erosion. Already many of the men who know most about animal breeding are in revolt; they are convinced this unnatural practice is bound to end in sterility and disaster.

The catalogue could be multiplied *ad infinitum*. The toll of disease is extraordinary and a matter of the utmost anxiety to the farmer. The public is not sufficiently aware of this unsatisfactory state of affairs. If these are the results of agricultural science, they are not encouraging and certainly are not impressive. They are undoubtedly a phenomenon of the last forty or fifty years and appear alongside of the modern use of artificial manures. This book asks the question whether we have here not things merely juxtaposed, but actual cause and result.

It is even more legitimate to ask what agricultural science would be at. It is a severe question, but one which imposes itself as a matter of public conscience, whether agricultural research in adopting the esoteric attitude, in putting itself above the public and above the farmer whom it professes to serve, in taking refuge in the abstruse heaven of the higher mathematics, has not subconsciously been trying to cover up what must be regarded as a period of ineptitude and of the most colossal failure. Authority has abandoned the task of illuminating the laws of Nature, has forfeited the position of the friendly judge, scarcely now ventures even to adopt the tone of the earnest advocate: it has sunk to the inferior and petty work of photographing the corpse—a truly menial and depressing task.

PART II

DISEASE IN PRESENT-DAY
FARMING AND GARDENING

A SIMPLE METHOD of estimating the success of any method of farming is to observe how it is affected by disease. If the soil is found to escape the two common ailments—erosion and the formation of alkali salts— which afflict cultivated land; if the crops raised are found to resist the various insect, fungous, and virus diseases; if the livestock breed normally and remain in good fettle; if the people who feed on such crops and livestock are vigorous, prolific, and more or less free from the many diseases from which mankind suffers; then the method of farming adopted is supported by the one unanswerable argument—success. It has passed the stiffest examination it can be made to undergo—it has yielded results comparable with those to be seen in the wayside hedges of this country of Great Britain. These strips closely resemble in their agriculture the primeval forest.

In our roadside hedges hardly a trace of the common diseases of the soil are to be seen; the wildings come into flower regularly every spring and early summer; there is no running out of the variety and no necessity to supply new and improved strains of seeds; one generation follows another century after century; the vegetable life of the hedgerow is to all intents and purposes eternal; there is very little plant disease. A similar story can be told of the birds and other animal life. The wayside hedge is, therefore, an example of successful soil management for all to see and study. It has stood the test of time.

In striking contrast to the picture of general health and well-being which has just been lightly sketched is the spectacle of widespread disease which has resulted from many of the methods of farming, and particularly the modern methods, which have so far been devised. Disease of one kind or another is the rule; robust health is the exception.

Let us, therefore, examine in some detail the generous dividends in the form of trouble with which Mother Earth has rewarded our methods of agriculture. The examples chosen have been largely taken from my own personal experience. They are arranged in their natural order starting with the diseases of the soil, then going on to the maladies of crops and livestock, and ending with the afflictions of *homo sapiens* himself.

PLATE V. BINS FOR COMPOSTING CANE TRASH AT SPRINGFIELD
ESTATE, NATAL

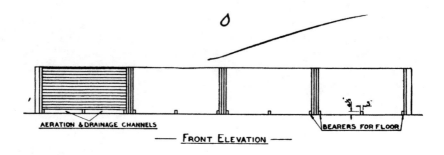

AERATION & DRAINAGE CHANNELS ⎯⎯ FRONT ELEVATION ⎯⎯ BEARERS FOR FLOOR

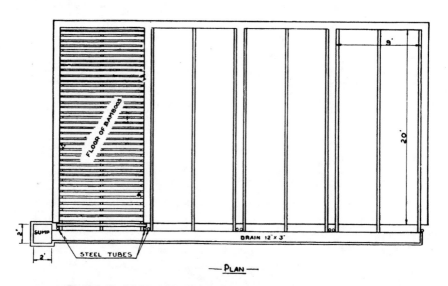

⎯ PLAN ⎯

PLATE VI. PLAN AND ELEVATION OF COMPOSTING BINS AT
SPRINGFIELD ESTATE, NATAL

7

SOME DISEASES OF THE SOIL

SOIL EROSION

Pᴇʀʜᴀᴘs ᴛʜᴇ ᴍᴏsᴛ widespread and the most important disease of the soil at the present time is soil erosion, a phase of infertility to which great attention is now being paid.

Soil erosion in the very mild form of denudation has been in operation since the beginning of time. It is one of the normal operations of Nature going on everywhere. The minute mineral particles which result from the decay of rocks find their way sooner or later to the ocean, but many may linger on the way, often for centuries, in the form of one of the constituents of fertile fields. This phenomenon can be observed in any river valley. The fringes of the catchment area are frequently uncultivated hills, through the thin soils of which the underlying rocks protrude. These are constantly weathered and in the process yield a continuous supply of minute mineral fragments in all stages of decomposition.

The slow rotting of exposed rock surfaces is only one of the forms of decay: the surfaces not exposed are also subject to change. The covering of soil is no protection to these underlying strata, but rather the reverse, because the soil water, containing carbon dioxide in solution, is constantly disintegrating the parent rock, first producing subsoil and then actual soil. In this way the constant supply of minerals—like phosphates, potash, and the trace elements needed by crops and livestock—are automatically transferred to the surface soil from the great mineral reservoir of the primary and secondary rocks. Simultaneously with these disintegration processes the normal decay of animal and vegetable remains on the surface of the soil is giving rise to the formation of humus.

All these processes combine to start up denudation. The fine soil particles of mineral origin, often mixed with fragments of humus, are gradually removed by rain, wind, snow, or ice to lower regions. Ultimately the rich valley lands are reached, where the accumulations may be many feet in thickness. One of the main duties of the streams and rivers which drain the valley is to transport these soil particles into the sea, where fresh land can be laid down. The process looked at as a whole is nothing more than Nature's method of the rotation, not of the crop, but of the soil itself. When the time comes for the new land to be enclosed and brought into cultivation, agriculture is born again. Such operations are well seen in England in Holbeach Marsh and similar areas round the Wash. From the time of the Romans to the present day new areas of fertile soil, which now fetch £100 an acre or even more, have been recreated from the uplands by the Welland, the Nene, and the Ouse. All this fertile land, perhaps the most valuable in England, is the result of two of the most widespread processes in Nature—weathering and denudation.

But Nature has devised a most effective brake. The nature of this retarding mechanism is of supreme importance, because it provides the key to the solution of the problem of soil erosion. Nature's control of the rate of denudation is to create the compound soil particle. The fragments of mineral matter derived from the weathering of rocks are combined by means of the specks of glue-like organic matter supplied mostly by the dead bodies of the soil bacteria which live on humus; as in a building made of bricks, some suitable cementing material is needed before the fragments of mineral matter in the soil can cohere. There must be sufficient of this cement of the right type always ready, so that when the mineral fragments come together a piece of glue is there at hand of a size corresponding to the minute areas of contact. This involves the constant production of large quantities of this bacterial cement. Provided, however, that we keep up the bacterial population of the land in any catchment area, the supplies of glue for making new compound soil particles and for repairing the old ones will be assured.

It will be seen from this how fundamentally important is the role of humus. It is the humus which feeds the bacterial life, which, so to say, glues the soil together and makes it effective. If the supply of glue is allowed to fall into arrears, the compound soil particles will soon lie about in ruins and so provide more raw material for speeding up the process of denudation. The mineral particles are thereby released and ready for their final journey by water to the sea to form new soil, or by wind to form a new dust bowl and so begin a new desert.

It is when the tempo of denudation is vastly accelerated by human agencies that a perfectly harmless natural process becomes transformed into a definite disease of the soil. The condition known as soil erosion—

a man-made disease—is then established. *It is, however, always preceded by infertility:* the inefficient, overworked, dying soil is at once removed by the operations of Nature and hustled towards the ocean, so that new land can be created and the rugged individualists—the bandits of agriculture—whose cursed thirst for profit is at the root of the mischief can be given a second chance. Nature is anxious to make a new and better start and naturally has no patience with the inefficient. Perhaps when the time comes for a new essay in farming, mankind will have learnt the great lesson—how to subordinate the profit motive to the sacred duty of handing over unimpaired to the next generation the heritage of a fertile soil. Soil erosion is nothing less than the outward and visible sign of the complete failure of a farming policy. The root causes of this failure are to be found in ourselves.

The damage already done by soil erosion all over the world, looked at in the mass, is very great and is rapidly increasing. The regional contributions to this destruction, however, vary widely. In some areas like north-western Europe, where most of the agricultural land is under a permanent or temporary cover crop (in the shape of grass or leys) and there is still a large area of woodland and forest, soil erosion is a minor factor in agriculture. In other regions like parts of North America, Africa, Australia, New Zealand, and the countries bordering the Mediterranean, where extensive deforestation has been practised and where almost uninterrupted cultivation has been the rule, large tracts of land once fertile have been almost completely destroyed.

The United States of America is perhaps the only country where anything in the nature of an accurate estimate of the damage done by erosion has been made. Theodore Roosevelt first warned the country as to its national importance. Then came the Great War with its high prices, which encouraged the wasteful exploitation of soil fertility on an unprecedented scale. A period of financial depression, a series of droughts and dust storms, emphasized the urgency of the salvage of agriculture. During Franklin Roosevelt's presidency soil conservation became a political and social problem of the first importance. In 1937 the condition and needs of the agricultural land of the United States of America were appraised. No less than 253,000,000 acres, or 61 per cent of the total area under crops, had either been completely or partly destroyed or had lost most of its fertility. Only 161,000,000 acres, or 39 per cent of the cultivated area, could be safely farmed by present methods. In less than a century the United States has, therefore, lost nearly

87

three-fifths of its agricultural capital. If the whole of the potential resources of the country could be utilized and the best possible practices introduced everywhere, about 447,466,000 acres could be brought into use—an area actually greater than the present crop land of 415,334,931 acres. The position, therefore, is not hopeless. It will, however, be very difficult, very expensive, and very time-consuming to restore the vast areas of eroded land even if money is no object and large amounts of manure are used and green-manure crops are ploughed under.

Such, in this great country, are the results of misuse of the land. The causes of this misuse include lack of individual knowledge of soil fertility on the part of the pioneers and their descendants; the traditional attitude which regarded the land as a source of profit; defects in farming systems, in tenancy, and finance—most mortgages contain no provisions for the maintenance of fertility; instability of agricultural production as carried out by millions of individuals, prices, and income, in contrast to industrial production carried on by a few large corporations. The need for maintaining a correct relation between industrial and agricultural production, so that both can develop in full swing on the basis of abundance, has only recently been understood. The country was so vast, its agricultural resources were so immense, that the profit seekers could operate undisturbed until soil fertility—the country's capital—began to vanish at an alarming rate.

The resources of the Government are now being called up to put the land in order. The magnitude of the effort, the mobilization of all available knowledge, the practical steps that are being taken to save what is left of the soil of the country and to help Nature to repair the damage already done are graphically set out in *Soils and Men,* the Year Book of the United States Department of Agriculture of 1938. This is perhaps the best local account of soil erosion which has yet appeared. The progress that has been made in recent years can be followed in *Soil Conservation,* a monthly periodical issued by the Soil Conservation Service of the United States Department of Agriculture, Washington, D. C.

The rapid exploitation of Africa was soon followed by soil erosion. In South Africa, a pastoral country, some of the best grazing areas are already semi-desert. The Orange Free State in 1879 was covered with rich grass, interspersed with reedy pools, where now only useless gullies are found. Towards the end of the nineteenth century, it began to be realized all over South Africa that serious over-stocking was taking place. In 1928 the Drought Investigation Commission reported that soil erosion was extending rapidly over many parts of the Union and that the eroded material was silting up reservoirs and rivers and causing a marked decrease in the underground water supplies. The cause of erosion was considered to be the reduction of vegetal cover brought about by incorrect veldt management—the concentration of stock in kraals, overstocking, and indiscriminate burning to obtain fresh autumn or winter grazing. In Basutoland, a normally well watered country, soil erosion is now the most immediately pressing administrative problem. The pressure of population has brought large areas under the plough and has intensified over-stocking on the remaining pasture. In Kenya the soil erosion problem has become serious during the last ten years, both in the native reserves and in the European areas. In the former, wealth depends on the possession of large flocks and herds; barter is carried on in terms of livestock; the bride price is almost universally paid in animals; numbers rather than quality are the rule. The natural consequence is overstocking, over-grazing, and the destruction of the natural covering of the soil. Soil erosion is the inevitable result. In the European areas, erosion is caused by long and continuous over-cropping without the adoption of measures to prevent the loss of soil and to maintain the humus content. Locusts have of late been responsible for greatly accelerated erosion; examples are to be seen when the combined effect of locusts and goats has resulted in the loss of a foot of surface soil in a single rainy season.

The countries bordering the Mediterranean provide striking examples of soil erosion, accompanied by the formation of deserts which are considered to be due to one main cause—the slow and continuous deforestation of the last 3,000 years. Originally well wooded, no forests are to be found in the Mediterranean region proper. Most of the original soil has been washed away by the sudden winter torrents. In North Africa the fertile cornfields which existed in Roman times are now desert. Ferrari in his book on woods and pastures refers to the changes in the soil and climate of Persia after its numerous and majestic parks were destroyed; the soil was transformed into sand; the climate became arid and suffocating; springs first decreased and then disappeared. Similar changes took place in Egypt when the forests were devastated; a decrease in rainfall and in soil fertility was accompanied by loss of uniformity in the climate. Palestine was once covered with valuable forests and fertile pastures and possessed a cool and moderate climate; to-day its mountains are denuded, its rivers are almost dry, and crop production is reduced to a minimum.

The above examples indicate the wide extent of soil erosion, the very serious damage that is being done, and the fundamental cause of the trouble—misuse of the land, resulting in the destruction of the compound soil particles. In dealing with the remedies which have been suggested and which are now being tried out, it is essential to envisage the real nature of the problem. It is nothing less than the repair of Nature's drainage system—the river—and of Nature's method of providing the countryside with a regular water supply. The catchment area of the river is the natural unit in erosion control. In devising this control we must restore the efficiency of the catchment area as a drain and also as a natural storage of water. Once this is accomplished, we shall hear very little about soil erosion.

Japan provides perhaps the best example of the control of soil erosion in a country with torrential rains, highly erodible soils, and a topography which renders the retention of the soil on steep slopes very difficult. Here erosion has been effectively held in check by methods adopted regardless of cost, for the reason that the alternative to their execution would be national disaster. The great danger from soil erosion in Japan is the deposition of soil debris from the steep mountain slopes on the rice fields below. The texture of the rice soils must be maintained so that the fields will hold water and allow of the minimum of through drainage. If such areas become covered with a deep layer of permeable soil, brought down by erosion from the hillsides, they would no longer hold water and rice cultivation—the mainstay of Japan's food supply—would be out of the question. For this reason the country has spent as much as ten times the capital value of eroding land on soil conservation work, mainly as an insurance for saving the valuable rice lands below. Thus, in 1925 the Tokyo Forestry Board spent 453 yen (£45) per acre in anti-erosion measures on a forest area valued at 40 yen per acre in order to save rice fields lower down valued at 240 to 300 yen per acre.

The dangers from erosion have been recognized in Japan for centuries and an exemplary technique has been developed for preventing them. It is now a definite part of national policy to maintain the upper regions of each catchment area under forest as the most economical and effective method of controlling flood waters and insuring the production of rice in the valleys. For many years erosion control measures have formed an important item in the national budget.

According to Lowdermilk (*Oriental Engineer,* March 1927), erosion control in Japan is like a game of chess. The forest engineer, after studying his eroding valley, makes his first move, locating and building one or more check dams. He waits to see what Nature's response is. This determines the next move which may be another dam or two, an increase in the former dam, or the construction of retaining side walls. After another pause for observation a further move is made and so on until erosion is checkmated. The operation of natural forces, such as sedimentation and re-vegetation, are guided and used to the best advantage to keep down costs and to obtain practical results. *No more is attempted than Nature has already done in the region.* By 1929 nearly 2,000,000 hectares of protection forests were used in erosion control. These forest areas do more than control erosion. They help the soil to absorb and retain large volumes of rain water and to release it slowly to the rivers and springs.

China, on the other hand, presents a very striking example of the evils which result from the inability of the administration to deal with the whole of a great drainage area as one unit. On the slopes of the upper reaches of the Yellow River extensive soil erosion is constantly going on. Every year the river transports over 2,000,000,000 tons of soil, sufficient to raise an area of 400 square miles by five feet. This is provided by the easily erodible loess soils of the upper reaches of the catchment area. Some of the mud is deposited in the river bed lower down, so that the embankments which contain the stream have constantly to be raised. Periodically the great river wins in this unequal contest and destructive inundations result. The labour expended on the embankments is lost, because the nature of the erosion problem as a whole has not been grasped, and the area drained by the Yellow River has not been studied and dealt with as a single organism. The difficulty now is the over-population of the upper reaches of the catchment area, which prevents afforestation and laying down to grass. Had the Chinese maintained effective control of the upper reaches—the real cause of the trouble— the erosion problem in all probability would have been solved long ago at a lesser cost in labour than that which has been devoted to the embankment of the river. China, unfortunately, does not stand alone in this matter. A number of other rivers, like the Mississippi, are suffering from overwork, followed by periodical floods as the result of the growth of soil erosion in the upper reaches.

Although the damage done by uncontrolled erosion all over the world is very great and the case for action needs no argument, nevertheless there is one factor on the credit side which has been overlooked. A considerable amount of new soil is being constantly produced by natural weathering agencies from the subsoil and the parent rock. This, when suitably conserved, will soon re-create large stretches of valuable land. One of the best regions for the study of his question is the black cotton soil of Central India which overlies the basalt. Here, although erosion is continuous, the soil does not often disappear altogether, for the reason that, as the upper layers are removed by rain, fresh soil is re-formed from below. The large amount of earth so produced is well seen in the Gwalior State, where the late ruler employed an irrigation officer, lent by the Government of India, to construct a number of embankments, each furnished with spillways, across many of the valleys, which had suffered so badly by uncontrolled rain wash in the past that they appeared to have no soil at all, the scrub vegetation just managing to survive in the crevices of the bare rock. How great is the annual formation of new soil, even in such unpromising circumstances, must be seen to be believed. In a few years the construction of embankments was followed by stretches of fertile land which soon carried fine crops of wheat. A brief illustrated account of the work done by the late Maharaja of Gwalior would be of great value at the moment for introducing a much needed note of optimism in the consideration of this soil erosion problem. Things are not quite so hopeless as they are often made to appear.

Why is the forest such an effective agent in the prevention of soil erosion? The forest does two things: (1) the trees and undergrowth break up the rainfall into fine spray and the litter on the ground further protects the soil from the impact of the descending water stream; (2) the residues of the trees and animal life met with in all woodlands are converted into humus, which is then absorbed by the soil underneath, increasing its porosity and water-holding power; the soil cover and the soil humus together prevent erosion and at the same time store large volumes of water. These factors—soil protection, soil porosity, and water retention—conferred by the living forest cover, provide the key to the solution of the soil erosion problem. All other purely mechanical remedies, such as terracing and drainage, are secondary matters, although, of course, important in their proper place.

The secret of soil conservation is thus seen to lie, first, in maintaining the soil cover in good condition to ensure that the rainfall is received on the surface in a proper manner with no disturbance of the soil below, and second, in conserving ample supplies of humus so that by means of the compound soil particles the water, when it has descended, is adequately absorbed and stored: as well might we expect a living creature to survive without its protective skin as to suppose that the earth can live without her proper covering. The forest has been cited as the preeminent example of these protective devices, for the leafage is thick and the ground litter abundant. In the absence of forest some form of grass cover is the natural protective agent which will for centuries often maintain the soil in good heart. Indeed, this device of the grass cover is far more efficient than might be supposed possible. The accumulations of humus under a grass carpet are often immense; they are, indeed, so extraordinary that they can be described as veritable mines of fertility. This is proved by the fact that an agriculture based on their spoliation can, in favourable circumstances, continue for many years before it fades out. But fade out it must if the humus is never restored. Williams (Timiriasev Academy, Moscow) regarded grass as the basis of all agricultural land utilization and the soil's chief weapon against the plundering instincts of humanity. He advanced the hypothesis that the decay of past civilizations was due to the wholesale ploughing up of grass necessitated by the increasing demands of civilization. His views are exerting a marked influence on soil conservation policy in the U.S.S.R. and indeed apply to many other countries.

Grass is a valuable factor in the correct design and construction of surface drains. Whenever possible these should be wide, very shallow, and completely grassed over. The run-off then drains away as a thin sheet of clear water, leaving all the soil particles behind. The grass is thereby automatically manured and yields abundant fodder. This simple device was put into practice at the Shahjahanpur Sugar Experiment Station in India. The earth service roads and paths were excavated so that the level was a few inches below that of the cultivated area. They were then grassed over, becoming very effective drains in the rainy season, carrying off the excess rainfall as clear water without any loss of soil.

If we regard erosion as the natural consequence of improper methods of agriculture and the catchment area of the river as the natural unit for the application of soil conservation methods, the various remedies available fall into their proper place. The upper reaches of each river system must be afforested; cover crops, including grass and leys, must be used to protect the arable surface whenever possible; the humus content of the soil must be increased and the crumb structure restored, so that each field can drink in its own rainfall; over-stocking and over-grazing must be prevented; simple mechanical methods for conserving the soil and regulating the run-off, like terracing, contour cultivation and contour drains, must be utilized. There is, of course, no single anti-erosion device which can be universally adopted. The problem must, in the nature of things, be a local one. Nevertheless, certain guiding principles exist which apply everywhere. First and foremost is the restoration and maintenance of the crumb structure of the soil, so that each acre of the catchment area can do its duty by absorbing its share of the rainfall.

THE FORMATION OF ALKALI LAND

When the land is continuously deprived of oxygen, the plant is soon unable to make use of the nourishment it contains: it becomes a dead instrument, from which no crop can draw anything. If left to itself, this condition of infertility is permanent.

In many parts of the tropics and sub-tropics agriculture is interfered with and even brought to an end because of the injury inflicted on the soil by accumulations of soluble salts composed of various mixtures of the sulphate, chloride, and carbonate of sodium. Such areas are known as alkali lands. When the alkali phase is still in the mild or incipient stage, crop production becomes difficult and care has to be taken to prevent matters from getting worse. When the condition is fully established, the soil dies; crop production is then out of the question. Alkali lands are common in Central Asia, India, Persia, Iraq, Egypt, North Africa, and the United States.

At one period it was supposed that alkali salts were the natural consequences of a light rainfall, insufficient to wash out of the land the salts which always form in it by progressive weathering of the rock powder, of which all soils largely consist. Hence alkali lands were considered to be a natural feature of arid tracts such as parts of north-west India, Iraq, and northern Africa, where the rainfall is very small. Such ideas of the origin and occurrence of alkali lands do not correspond with the facts and are quite misleading. The rainfall of the Province of Oudh in India, for example, where large stretches of alkali lands naturally occur, is certainly adequate to dissolve the comparatively small quantities of soluble salts found in these infertile areas, if their removal were a question of sufficient water only. In North Bihar the average rainfall in the submontane tracts where large alkali patches are common is about fifty to sixty inches a year. Arid conditions, therefore, are not essential for the production of alkali soils; heavy rainfall does not always remove them.

What is a necessary condition is impermeability. In India, whenever the land loses its porosity by the constant surface irrigation of stiff soils with a tendency to impermeability, by the accumulation of stagnant subsoil water, or through some interference with surface drainage, alkali salts sooner or later appear. Almost any agency, even over-cultivation or over-stimulation by means of artificial manures, both of which oxidize the organic matter and slowly destroy the crumb structure, will produce alkali land. In the neighbourhood of Pusa in North Bihar old roads and the sites of bamboo clumps and of certain trees, such as the tamarind (*Tamarindus indica* L.) and the pipul (*Ficus religiosa* L.) always give rise to alkali patches when they are brought into cultivation. The densely packed soil of such areas invariably shows the bluish-green markings which are associated with the activities of those soil organisms existing in badly aerated soils without a supply of free oxygen. A few inches below the alkali patches which occur on the stiff, loess soils of the Quetta valley, similar bluish-green and brown markings always occur. In the alkali zone in North Bihar wells have always to be left open to the air, otherwise the water is contaminated by sulphuretted hydrogen, thereby indicating a well-marked, reductive phase in the deeper layers. In a subsoil drainage experiment on the black soils of the Nira valley in Bombay, where perennial irrigation was followed by the formation of alkali land, Mann and Tamhane found that the salt water which ran out of these drains soon smelt strongly of sulphuretted hydrogen and a white deposit of sulphur was formed at the mouth of each drain, proving how strong were the reducing actions in this soil. Here the reductive phase in alkali formation was unconsciously demonstrated in an area where alkali salts were unknown until the land was waterlogged by over-irrigation and the oxygen supply of the soil was restricted.

The view that the origin of alkali land is bound up with defective soil aeration is supported by the recent work on the origin of salt water lakes in Siberia. In Lake Szira-Kul between Bateni and the mountain range of Kizill Kaya, Ossendowski observed in the black ooze taken from the bottom of the lake and in the water a certain distance from the surface an immense network of colonies of sulphur bacilli, which gave off large quantities of sulphuretted hydrogen and so destroyed practically all the fish in this lake. The great water basins in central Asia are being metamorphosed in a similar way into useless reservoirs of salt water, smelling strongly of hydrogen sulphide. In the limans near Odessa and in portions of the Black Sea a similar process is taking place. The fish, sensing the change, are slowly leaving this sea as the layers of water, poisoned by sulphuretted hydrogen, are gradually rising towards the surface. The death of the lakes scattered over the immense plains of Asia and the destruction of the impermeable soils of this continent from alkali salt formation are both due to the same primary cause—intense oxygen starvation. In the instances just mentioned this oxygen starvation occurs naturally; in other cases it follows perennial irrigation.

Every possible gradation in alkali land is met with. Minute quantities of alkali salts in the soil have no injurious effect on crops or on the soil organisms. It is only when the proportion increases beyond a certain limit that they first interfere with growth and finally prevent it altogether. Leguminous crops are particularly sensitive to alkali, especially when this contains carbonate of soda. The action of alkali salts on the plant is a physical one and depends on the osmotic pressure of solutions, which increases with the amount of the dissolved substance. For water to pass readily from the soil into the roots of plants, the osmotic pressure of the cells of the root must be considerably greater than that of the soil solution outside. When the soil solution becomes stronger than that of the cells, water passes backwards from the roots to the soil and the crops dry up. This state of affairs inevitably occurs when the soil becomes charged with alkali salts beyond a certain point. The crops are then unable to take up water and death results. The roots behave like a plump strawberry when placed in a strong solution of sugar; like the strawberry they shrink in size because they have lost water to the stronger solution outside. Too much salt in the water, therefore, makes irrigation water useless and destroys the canal as a commercial proposition.

The reaction of the crop to the first stages in alkali production is interesting. For twenty years at Pusa and eight years in the Quetta valley I had to farm land, some of which hovered, as it were, on the verge of alkali. The first indication of the condition is a darkening of the foliage and the slowing down of growth. Attention to soil aeration, to the supply of organic matter, and to the use of deep-rooting crops like lucerne and

96

the pigeon pea, which break up the subsoil, soon set matters right. Disregard of Nature's danger signals, however, leads to trouble—a definite alkali patch is formed. When cotton is grown under canal irrigation on the alluvial soils of the Punjab, the reaction of the plant to incipient alkali is first shown by the failure to set seed, on account of the fact that the anther, the most sensitive portion of the flower, fails to function and to liberate its pollen. The cotton plant naturally finds it difficult to obtain from mild alkali soil all the water it needs—this shortage is instantly reflected in the breakdown of the floral mechanism.

Is the alkali condition confined to the tropics and sub-tropics? May it not, under certain circumstances, occur in temperate regions such as north-western Europe? Is it a factor in the sandy soils of Wareham in Dorsetshire recently investigated by Professor Neilson-Jones and Dr. Rayner? It is impossible at the moment to answer these questions till the soil studies of the future consider the biological activities in relation to the physical and chemical factors as well as to the season. They may not have reached the grade of decay known as alkali land, but they are starved of oxygen, all the conditions needed for the establishment of the anaerobic and semi-anaerobic state being present. This is made clear by the readiness with which they respond to any improvement in surface and subsoil drainage, as well as to sub-soiling. Soil conditions must be looked at as a living and changing system and not merely as something static and stable. The soils of the north temperate zone, for example, often suffer from poor soil aeration. Moreover, many of the soil profiles exhibit the blue and red markings so common under alkali patches, as well as bands of humus which must have been originally formed near the surface, then carried in solution and afterwards precipitated. The soil organisms, which reduce compounds containing sulphur to sulphuretted hydrogen, are known to exist in these soils. All facts point to the necessity for further work so as to provide a clear answer to the above mentioned questions, while from the practical point of view there is an immense field for improvement, especially by means of sub-soiling, over many areas which are now allowed to continue in a very unsatisfactory state. The problem of soil aeration is by no means, therefore, confined to the tropics, and it behooves the pioneers of farming in the temperate countries to turn an immediate attention to the various fairly simple devices by which very great, and above all, permanent improvements could be effected.

The stages in the development of the alkali condition are somewhat as follows. The first condition is an impermeable soil. Such soils—the *usar* plains of northern India for example—occur naturally where the climatic condition favour those biological and physical factors which destroy the soil structure by disintegrating the compound particles into their ultimate units. These latter are so extremely minute and so uniform in size that they form with water a mixture possessing some of the properties of colloids which, when dry, pack into a hard, dry mass, practically impenetrable to water and very difficult to break up. Such soils are very old. They have always been impermeable and have never come into cultivation.

In addition to the alkali tracts which occur naturally, a number are in course of formation as the result of errors in soil management, the chief of which are as follows:

(*a*) The excessive use of irrigation water: this gradually destroys the binding power of the organic cementing matter which glues the soil particles together, and further displaces the soil air. Anaerobic changes, indicated by blue and brownish markings, first occur in the lower layers and finally lead to the death of the soil. It is this slow destruction of the living soil that must be prevented if the existing schemes of perennial irrigation are to survive. The process is taking place before our eyes to-day in the Canal Colonies of India, where irrigation is loosely controlled.

(*b*) Over-cultivation without due attention to the replenishment of humus: in those continental areas like the Indo-Gangetic plain, where the risk of alkali is greatest, the normal soils contain only a small reserve of humus, because the biological processes which consume organic matter are very intense at certain seasons, due to sudden changes from low to very high temperatures and from intensely dry weather to periods of moist, tropical conditions. Accumulations of organic matter such as occur in temperate zones are impossible. There is, therefore, a very small margin of safety. The slightest errors in soil management will not only destroy the small reserve of humus in the soil, but also the organic cement on which the compound soil particles and the crumb structure depend. The result is impermeability, the first stage in the formation of alkali salts. The inhabitants of these areas through the centuries have followed methods of cultivation which are perfectly adapted to preserve the safety margin, but there is a tendency on the part of the short-sighted Western scientist to teach them so-called techniques of stimulating crop production which are highly dangerous from this point of view.

98

One suggestion that is constantly being put forward is the introduction into the Indo-Gangetic plain of artificial fertilizers like sulphate of ammonia. This would soon lead to catastrophe.

(c) The use of artificial manures, particularly sulphate of ammonia: even where there is a large safety margin, i.e. a large reserve of humus, such dressings do untold harm. The presence of additional combined nitrogen in an easily assimilable form stimulates the growth of fungi and other organisms which, in the search for the organic matter needed for energy and for building up microbial tissue, use up first the reserve of soil humus and then the more resistant organic matter which cements the soil particles. This glue is not affected by the processes going on in a normally cultivated soil, but it cannot withstand the same processes when stimulated by dressings of artificial manures.

Alkali land, therefore, starts with a soil in which the oxygen supply is permanently cut off. Matters then go from bad to worse very rapidly. All the oxidation factors which are essential for maintaining a healthy soil cease. A new soil flora—composed of anaerobic organisms which obtain their oxygen from the sub-stratum—is established. A reduction phase ensues. The easiest source of oxygen—the nitrates—is soon exhausted. The organic matter then undergoes anaerobic fermentation. Sulphuretted hydrogen is produced as the soil dies, just as in the lakes of central Asia. The final result of the chemical changes that take place is the accumulation of the soluble salts of alkali land—the sulphate, chloride, and carbonate of sodium. When these salts are present in injurious amounts, they appear on the surface in the form of snow-white and brownish-black incrustations. The former (white alkali) consists largely of the sulphate and chloride of sodium, and the latter (the dreaded black alkali) contains sodium carbonate in addition and owes its dark colour to the fact that this salt is able to dissolve the organic matter in the soil and produce physical conditions which render drainage impossible. According to Hilgard, sodium carbonate is formed from the sulphate and chloride in the presence of carbon dioxide and water. The action is reversed in the presence of oxygen. Subsequent investigations have modified this view and have shown that the formation of sodium carbonate in soil takes place in stages. The appearance of this salt always marks the end of the chapter. The soil is dead. Reclamation then becomes difficult on account of the physical conditions set up by these alkali salts and the dissolved organic matter.

The occurrence of alkali land, as would be expected from its origin,

99

is extremely irregular. When ordinary alluvial soils like those of the Punjab and Sind are brought under perennial irrigation, small patches of alkali first appear where the soil is heavy; on stiffer areas the patches are large and tend to run together. On open, permeable stretches, on the other hand, there is no alkali. In tracts like the western districts of the United Provinces, where irrigation has been the rule for a long period, zones of well aerated land carrying fine irrigated crops occur alongside the barren alkali tracts. Iraq also furnishes interesting examples of the connection between alkali and poor soil aeration. Intensive cultivation under irrigation is only met with in that country where the soils are permeable and the natural drainage is good. Where the drainage and aeration are poor the alkali condition at once becomes acute. There are, of course, a number of irrigation schemes, such as the staircase cultivation of the Hunzas in north-west India and of Peru, where the land has been continually watered from time immemorial without any development of alkali salts. In Italy and Switzerland perennial irrigation has been practised for long periods without harm to the soil. In all such cases, however, careful attention has been paid to drainage and aeration and to the maintenance of humus; the soil processes have been confined by Nature or by man to the oxidative phase; the cement of the compound particles has been protected by keeping up a sufficiency of organic matter.

The theory of the reclamation of alkali land is very simple. All that is needed, after treating the soil with sufficient gypsum (which transforms the sodium clays into calcium clays), is to wash out the soluble salts, to add organic matter, and then to farm the land properly. Such reclaimed soils are then exceedingly fertile and remain so. If sufficient water is available, it is sometimes possible to reclaim alkali soils by washing only. I once confirmed this. The berm of a raised water channel at the Quetta Experiment Station was faced with rather heavy soil from an alkali patch. The constant passage of the irrigation water down the water channel soon removed the alkali salts. This soil then produced some of the heaviest crops of grass I have ever seen in the tropics. When, however, the attempt is made to reclaim alkali areas on a field scale by flooding and draining, difficulties at once arise unless steps are taken first to replace all the sodium in the soil complex by calcium and then to prevent the further formation of sodium clays. Even when these reclamation methods succeed, the cost is always considerable; it soon becomes prohibitive; the game is not worth the candle. The removal of alkali salts is only the first step; large quantities of organic matter are then needed; adequate soil aeration must be provided; the greatest care must be taken to preserve these reclaimed soils and to see that no reversion to the alkali condition occurs. It is exceedingly easy under canal irrigation to create alkali salts on certain areas. It is exceedingly difficult to reverse the process and to transform alkali land back again into a fertile soil.

An interesting development in the reclamation of alkali soils has recently taken place at the Coleyana Estate in the Montgomery District of the Punjab. The method adopted is a first-rate pointer to the right way of solving this or any other agricultural problem. It consists in a clever diagnosis of natural processes and an ingenious adaptation of them to attain the wished-for end. Nature is made, as it were, to retrace certain steps so as to re-establish more desirable soil conditions; she is asked to undo her own work. On the Coleyana Estate Colonel Sir Edward Hearle Cole, C.B., C.M.G., first removes the accumulations of alkali salts from the surface, then ploughs them up and plants *dhup* grass (*Cynodon dactylon,* Pers.) which is grazed as heavily as possible by sheep and cattle for some eighteen months to two years. The turf is then killed by a turnover plough followed by a fallow during the hot season (May and June). The land is then prepared for a green-manure crop, followed by a couple of wheat crops in succession, and then put into lucerne or cotton. The great thing in this reclamation work is to scrape off all alkali salts as they appear, remove them from the land, and use the minimum irrigation water for the establishment and mainte-nance of the crop of grass. The underground stems and roots of the grass then aerate the heavy soil: the sheet-composting of the turf and the droppings of the livestock create the large quantities of humus needed to get this heavy land into condition for wheat, cotton, and lucerne. Sir Edward is now making a point of never leaving such reclaimed land uncovered so as to make the fullest use of the energy of sunlight in creating vegetable matter, which ultimately gets con-verted into humus. He also takes advantage of deep-rooting plants such as chicory, lucerne, and *arhar* (*Cajanus indicus,* Spreng.) for breaking up the subsoil and is a firm believer in the principles set out in *The Clifton Park System of Farming.* In this way, areas once ruined by alkali salts are now producing crops of wheat up to 1,600 lb. to the acre. This is, perhaps, the simplest and easiest method of reclaiming alkali soils that has yet been devised. It makes the crop itself do most of the work. (*Indian Farming,* I, 1940, p. 280.)

A further development of the Coleyana method of reclaiming alkali land suggests itself. When the grass crop is ploughed up, it might be worth while to sub-soil the land to a depth of fifteen to eighteen inches four feet apart, using a caterpillar tractor and a Ransomes sub-soiler. This would shatter the deeper soil layers, provide abundant aeration, and prepare the land for the succeeding crops.

Nature has provided, in the shape of alkali salts, a very effective censorship for all schemes of perennial irrigation. The conquest of the desert by the canal by no means depends on the mere provision of water and arrangements for the periodical flooding of the surface. This is only one of the factors of the problem. The water must be used in such a manner and the soil management must be such that the fertility of the soil is maintained intact. There is obviously no point in creating at vast expense a Canal Colony and producing crops for a generation or two, followed by a permanent desert of alkali land. Such an achievement merely provides another example of agricultural banditry. It must always be remembered that the ancient irrigators never developed any efficient method of perennial irrigation, but were content with the basin system,[1] a device by which irrigation and soil aeration can be combined. In his studies on irrigation and drainage, King concludes an interesting discussion of this question in the following words which deserve the fullest consideration on the part of the irrigation authorities all over the world:

"It is a noteworthy fact that the excessive development of alkalis in India, as well as in Egypt and California, is the result of irrigation practices modern in their origin and modes and instituted by people lacking in the traditions of the ancient irrigators, who had worked these same lands thousands of years before. The alkali lands of to-day, in their intense form, are of modern origin, due to practices which are evidently inadmissible, and which in all probability were known to be so by the people whom our modern civilization has supplanted."

These words should be studied by all who are concerned with the extension of irrigation schemes. The unwise pursuance of such schemes with a view to the immediate production of easily grown crops without the lasting maintenance of fertility can only end in the regular suffocation of precious tracts of the earth's surface.

[1] The land is embanked; watered once; when dry enough it is cultivated and sown. In this way water can be provided without any interference with soil aeration.

8

THE DISEASES OF CROPS

Disease in crops manifests itself in a great variety of ways. Troubles due to parasitic fungi and insects are by far the most common. Many of these troubles have occurred from time to time all through the ages and are by no means confined to modern farming. In recent years attention has been paid to a number of other diseases, such as those due to eelworm, to virus, and to the loss of the power of the plant to reproduce itself. The varieties of our cultivated crops nowadays show a great tendency to run out and to become unremunerative. This weakness, which might be described as varietal-erosion or species-erosion, has to be countered by the creation of a constant stream of new varieties obtained either by plant breeding methods or by importation from other localities. Besides the many cases of running out, failure to set seed is also due to unfavourable soil conditions, the removal of which puts an end to the trouble.

The great attention now devoted to disease will be clear from the operations of the Empire Cotton Growing Corporation, a State-aided body incorporated by Royal Charter on 1st November 1921 for the development of cotton production in the Empire. Among the many activities of this Corporation is the publication of the *Empire Cotton Growing Review,* a feature of which are the notes on current literature. During the six years before the war, 1934–9, these abstracts of papers on cotton research cover 964 pages of print, of which no less than 223, i.e. 23 per cent, deal with the diseases of cotton. These figures roughly correspond with the way the money contributed all over the world for the production, improvement, and testing of new cottons is spent. Some quarter of the technical staff engaged in this work devote their whole time to the study of the diseases of the cotton plant.

That something must be wrong with the production of cotton throughout the Empire and indeed throughout the world is suggested by a comparison between the above alarming figures and my own experience at the Institute of Plant Industry at Indore in Central India, at which research centre cotton was the principal crop. Between the years 1924 and 1931 cotton disease at Indore was to all intents and purposes negligible. I can recall only one case of wilt on some half dozen plants in a waterlogged corner of a field in a year of exceptionally high rainfall. The cotton plant in India always impressed me as a robust grower capable of standing up well to adverse soil and weather conditions. The examples of disease I came across in my many tours always seemed to be a consequence of bad farming, all capable of elimination by improved methods of agriculture.

As my adventures in research began in the West Indies in 1899 as a mycologist, I have naturally followed very closely the subsequent work on the various diseases of crops and have always been interested in the many outbreaks of these troubles which have occurred all over the world. Since 1905 I have been in a position to grow crops myself and thus have been able to test the validity of the principles on which the conventional methods of disease control are based. Perhaps the simplest way of dealing with these experiences, observations, and reflections will be crop by crop.

In perusing the following pages one thing will strike the reader forcibly. I have found it impossible to separate the disease from the growing crop. The study of plant diseases for their own sake is proving an increasingly intricate game, to which modern scientists have devoted many wasted hours. Such studies would be amusing if they were not tragic, for no disease in plant, animal, or man can properly be viewed unless it is looked on as an interference with, or to speak more plainly, as the distortion or negation of that positive aspect of the growing organism which we call health.

Consequently it is essential to conceive of the plant, for instance, as a living and growing thing, flourishing in certain conditions but wilting or perishing in other conditions; in any discussion of plant disease the right and the wrong methods of growing the crop are not simply the background to the argument, they are its very substance: to investigate plant diseases without a first-hand experience of growing the plant is to play Hamlet without the Prince of Denmark.

SUGAR-CANE

While in the West Indies (1899–1902) I devoted much attention to the fungous diseases of sugar-cane, but only succeeded in writing a few routine papers on the subject, all of no particular importance. Some twenty-five years later at Indore I grew a number of excellent crops of cane and converted them into crude sugar, both of which proceedings won the approval of the local Indian population. This experience brought out one of the weaknesses in present-day research. Between the years 1899 and 1902 I could only write technical papers on the diseases of the cane, as I had no opportunity of growing the crop or of manu-facturing it into sugar. I was then in the strait-jacket stage of my career. It was not till a quarter of a century later in another continent that the chance came to grow sugar-cane, to the study of whose diseases I had devoted so much attention. It is safe to say that, had these periods been reversed, my papers on the fungous diseases of cane would have made very different reading.

The methods adopted in growing sugar-cane on the black cotton soils at Indore were a copy of those devised by the late Mr. George Clarke, C.I.E., at the Shahjahanpur Experiment Station and described in detail in Chapter XIV of *An Agricultural Testament*. The crop is planted in shallow trenches, two feet wide, four feet from centre to centre, the soil from each trench being removed to a depth of six inches and piled on the two-foot space left between each two trenches, the whole making a series of ridges as illustrated in Fig. 1.

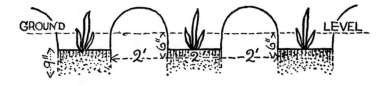

FIG. 1. Trench System at Indore

As soon as the trenches are made in November, they are dug to a further depth of six inches and compost is thoroughly mixed with the soil of the floor of the trenches, which are then watered, cultivated when dry enough, and allowed to remain till planting time in February. In this way the soil in which the cuttings are to be planted is given time to prepare the food materials needed when growth begins. After plant-ing and watering, the surface soil is lightly cultivated to prevent drying out. Afterwards four or five waterings are given, each followed by sur-face cultivation, which carry on the crop during the hot season till the break of the rains in June, when no further irrigation is needed.

105

When the young canes are about two feet high and are tillering vigorously, the trenches are gradually filled in, beginning about the middle of May and completing the operation by the middle of June, when the earthing up of the canes commences. This operation is completed about the middle of July (Fig. 2).

FIG. 2. Earthing up sugar-cane at Shahjahanpur, 10th July 1919

One of the consequences of filling in the trenches and of earthing up canes grown in fertile soil is the copious development of fungi, which are plainly visible as threads of white mycelium all through the soil of the ridges and particularly round the active roots. I saw these for the first time at the Manjri sugar-cane farm near Poona about 1920 and the same thing was frequently observed at Shahjahanpur. No one suspected then that this fungous development could be explained by the fact that the sugar-cane is a mycorrhiza former and that we were observing the first stage of an important symbiosis between the fungi living on the humus in the soil and the sap of the sugar-cane. The provision of all the factors needed for this association—humus, soil aeration, moisture, and a constant supply of fresh, active roots from the lower nodes of the canes as the earthing-up process proceeds—explains why such good results have always followed the Shahjahanpur method of growing the cane and why the crops are so healthy. When grown on the flat under monsoon conditions, want of soil aeration and want of a constant supply of fresh roots would always be limiting factors in the full establishment of the mycorrhizal association.

As at Shahjahanpur, the operation of earthing up the canes served four purposes: (1) the succession of new roots arising from the lower nodes, thoroughly combed the highly aerated and fertile soil of the ridges; (2) the conditions suitable for the constant development of the mycorrhizal association were provided; (3) the standing power of the

106

canes during the rains was vastly improved, and (4) the excessive development of colloids in the surface soil was prevented. When this earthing up is omitted, a heavy crop of cane is liable to be levelled by the monsoon gales; crops which fall down during the rains do not ripen properly, do not give either the maximum yield of sugar or the much-prized, light-coloured product.

The operation of earthing up left deep drains between the rows of cane. It was essential, as at Shahjahanpur, to arrange that these drains were suitably connected with the ditches which carried off the surplus monsoon rainfall, so that no waterlogging of the area under cane occurred.

At Indore the Shahjahanpur results were repeated. The intensive cultivation of a suitable variety (POJ 213 and Coimbatore 213), proper soil aeration, good surface drainage, and an adequate supply of organic matter produced very fine yields of cane, free from fungous and virus diseases and exceptionally good samples of crude sugar (*gur*). The yields were not quite up to the Shahjahanpur standard, because it takes some years to work up the black soils to the highest pitch of fertility on account of the physical character of these heavy soils, but I am convinced that this was only a matter of perseverance. Unfortunately the time of retirement came before I could achieve the full results, but the remarkable yields obtained in the first three years left no doubt in my mind of the final result. There is no question but that the way to grow cane is the Shahjahanpur method, which should be adopted all over the world, particularly for raising the plant material.

No fungous or virus diseases were observed at Indore. The growth of cane and the ripening process were almost ideal. But not quite. It was noticed that the length of the nodes formed under irrigation during the hot season was rather short. Some factor seemed to be retarding growth during this period. At the time I put this down to the fact that the land under cane had only just been brought under irrigation and that insufficient time had been allowed to get these fields into that high state of fertility so essential when ordinary, rain-fed, black soils are converted into well-irrigated land. As a rule this takes five years in Central India. This retardation in growth during the hot season was accompanied by a very mild attack of the moth borer (*Diatrea saccharalis*), which lays its eggs in clusters on the under-side of the leaves and is followed by the destruction of the young shoots invaded by the caterpillars. Only a few shoots were destroyed; nothing was done to check the moth. As soon, however, as the rains broke, this pest disappeared of its own accord and no further damage occurred. Obviously some factor was operating during the hot season which altered the sap and lowered the resistance of the cane. I suspected at the time that the soil was not sufficiently fertile and did not contain sufficient humus for supplying the young growing cane with all the water it needed, and that this very minor trouble would disappear when the irrigated area was got into really good fettle. This is obviously a matter calling for detailed investigation.

At Indore the only manure used in raising the cane crop was compost. At Shahjahanpur the canes were grown on green-manure supplemented by a light dressing of cattle manure applied to the land before the green crop was sown. The only examples of organic manuring in commercial cane growing I have been able to discover are in Mauritius, where livestock are kept solely for their manure, which is used to break down cane trash into a rough form of compost. Thus at the Benares estate the residues of 140 cattle are converted into 1,500 tons of compost at a total cost of 6s. 6d. a ton. At Mon Trésor estate 5,000 tons of compost were made at a similar cost from the residues of 300 cattle and 500 sheep and goats. Further details of this organic manuring in Mauritius are to be found in a paper by G. C. Dymond reprinted in the *News-Letter on Compost*, No. 7, October 1943, p. 44.

In recent years another type of sugar-cane disease—virus—has assumed considerable importance. If virus is nothing more than a condition caused by imperfectly synthesized protein, aggravated by the use of artificials like sulphate of ammonia in place of humus, it would follow that a drastic alteration in manuring might remove the virus condition and restore health. In Natal this has been accomplished. Mr. G. C. Dymond found that when Uba canes, attacked by streak disease (a virus trouble), were manured with compost and the process was repeated for a year or two, the crop threw off the disease and grew normally. The restoration of health was accompanied by the establishment of the mycorrhizal association, which was absent in the cases of streak disease examined.

Dymond's discovery that freshly prepared compost not only restores virus-infected canes to health, but also re-establishes the mycorrhizal association, is of great importance in the future studies of cane diseases. The first step in such inquiries should be to examine the mycorrhizal status of the affected plants and then to restore it by growing cuttings of

108

the diseased plants in heavily composted soil. In all probability the disease will disappear. Steps should then be taken to apply this knowledge on a field scale and then to see whether such crops can be infected by disease. If, as is most probable, no infection takes place, then the cause of the trouble—bad farming—has been established, as well as the remedy—freshly prepared humus.

The next step will be to see how many of the fungous, insect, and virus diseases of the cane survive the Shahjahanpur methods of cane growing. This at least is certain—the number will be few, perhaps none. In this way sugar-cane pests can be used as agricultural censors; their prevention will tune up practice; mycologists and entomologists will then become active and useful agents in development.

Intimately bound up with the prevention of cane diseases is the maintenance of the variety. As has already been pointed out (p. 10), the kinds of cane grown in the East have lasted for many centuries; on the modern sugar plantations a constant stream of new kinds has to be created. The prevention of this deterioration would seem to be bound up with the prevention of disease—the maintenance without any sign of progressive deterioration in the synthesis of protein. This is accomplished in the indigenous sugar industry of India by the use of cattle manure and the restriction of the cuttings used in planting to the joint immediately below the cane tops. These are buried at harvest time and carefully kept till the new field is planted. Commercial sugar estates might copy this well-tried practice and so save the time and money expended in testing a constant stream of new canes.

COFFEE

In the course of my travels I have seen something of coffee cultivation —in the West Indies, in various parts of India, and in the coffee-growing areas of Africa. I also visited in 1908 and again in 1938 the eroded areas in the centre of Ceylon which were devoted to coffee till the well-known rust fungus—*Hemileia vastatrix*—destroyed the plantations wholesale and caused them to be planted in tea. In all this two things impressed me very much: (1) the marked response of the coffee bush to forest soils rich in humus, and (2) the poor growth seen on areas suffering from erosion. On reconsidering in 1938 the original accounts of the great fungous epidemic in Ceylon some sixty years before, it appeared to me that the loss of the fertile top soil by erosion and the inadequate provision of fresh supplies of humus were ample reasons why this coffee disease had put an end to the industry. This surmise was strengthened by the establishment of the fact that coffee is a mycorrhiza former. This point is referred to in the following extract from my report dated 18th April 1938 on a visit to the tea estates in India and Ceylon:

"In view of the results obtained on the coffee estates in Kenya and Tanganyika with compost, it was expected that mycorrhiza would be found in this crop. Unfortunately my tour did not include any coffee estates where the Indore Process had been adopted. Three samples of surface roots, however, were collected.

"The first was taken from stray coffee plants growing on the roadside on unmanured land under grass at Dholai (Cachar, Assam). As was expected, Dr. Rayner found no trace of mycorrhiza in these root samples.

"Two more promising samples were collected at Talliar (High Range, Travancore), one from a nursery, the other from established coffee. In both cases the soil contained forest humus and in both Dr. Rayner found endotrophic fungous infection of the same type as that described in tea, but confined to the older roots and sporadic in distribution.

"The evidence, although incomplete and fragmentary, nevertheless points to mycorrhiza being as important a factor in coffee cultivation as it is proving in tea."

These observations were confirmed and amplified by the examination of material sent from Costa Rica by Señor Don Mariano Montealegre. There is no doubt that coffee, like tea and cacao, is a mycorrhiza former.

The fact that coffee is a mycorrhiza former is of considerable significance in the future cultivation of this crop. The humus in the soil and the sap of the plant are in intimate contact by means of this natural mechanism. Obviously, therefore, if coffee of the highest quality is to be produced and if the plants are to withstand disease, the first condition of success in coffee cultivation is the provision of properly made humus. This naturally involves some form of mixed farming so that an ample supply of urine and dung is available on the spot. Pigs, buffaloes, and cattle will probably be the best agents for this purpose. The day, therefore, may not be far distant when the coffee estates will be partly devoted to livestock, which will automatically cancel out the present expenditure on artificial manures and insecticides, and do much to raise the yield per acre and also improve the quality—a matter of supreme importance in this crop.

One illuminating consequence of the devastating epidemic of coffee leaf disease in Ceylon impressed me during my tours in the island in 1908 and thirty years later in 1938. The many planters I met not only had not forgotten this visitation, but were still labouring under the thraldom of fear of the parasite. When I suggested that fungous and insect diseases are the direct consequence of mistakes in crop production and should, therefore, be regarded as friendly professors of agriculture provided by Nature free of charge for our instruction, I found myself up against a solid armour-plate of fear. Diseases, like erosion, were things which had to be studied by specialists and then tackled by direct action.

Under these unpromising conditions I did not pursue the subject and go on to suggest that *Hemileia vastatrix* would prove most useful in another way. This disease of the coffee plant might well be used not only to teach us how to grow coffee properly, but also in reference to another crop—the tea plant. A few coffee plants, established here and there among the tea, would tell us whether the soils of Ceylon had been sufficiently restored to fertility by the anti-erosion methods undertaken, by the planting of adequate shade, and above all by the practice of systematically converting all vegetable and animal residues into humus. They could do this without any soil analyses or other laboratory tests by simply withstanding the onset of the leaf disease or by succumbing to it; where the disease appeared, we should know that the soil still lacked fertility; when it was absent, we should be able to be satisfied with the measures taken.

Such a device would be very simple. It would be efficient because it would be using Nature's own agencies in testing conditions. Why should we not make use of so excellent and so inexpensive a method? The Ceylon tea planter should look on coffee *and the diseases it carries* as one of his best, his most willing, and his most reliable assistants.

TEA

Although a number of insect and fungous diseases have been reported on the tea plant, nevertheless the total damage done by these pests is not excessive. Nothing like the coffee leaf disease of Ceylon, which in a few years destroyed the plantations wholesale, has been reported in the case of tea. Indeed in Ceylon, as has already been stated, tea replaced coffee on the partially eroded soils, a fact which suggests that the tea bush is exceptionally hardy and robust. This view is confirmed by the behaviour of this species under cultivation. The plants are constantly plucked and so deprived of those portions of their foliage richest in food materials; every few years the bushes are heavily pruned, after which they have to re-create themselves; in China a tea plantation lasts a century or more. Only a very vigorous bush could endure such treatment for so long.

It would follow from all these considerations that the struggle between the host and the parasite might easily result in the victory of the former, if the tea plant were given a little assistance. It might then be easy to reduce the damage done by pests to something quite insignificant.

Can the tea plant itself throw any light on this question of natural resistance to disease? Has the tea bush anything to say about the assistance it needs to vanquish the various insect and fungous pests always ready to attack it? If so, its representations must be carefully studied and if possible implemented. The plant or the animal will answer most queries about its needs if the questions are properly posed. The wise farmer, planter, or gardener always deals with such responses with sympathy and respect.

The tea plant has very recently delivered a most emphatic message on the cause of disease and its prevention which is certain to interest many readers in no way connected with the tea industry. The story I have to tell began in 1933 when I interested myself in the career of Dr. C. R. Harler (who had just been retrenched when the Tocklai Research Station, maintained by the Indian Tea Association, was reorganized in that year). I consoled him for his temporary loss of employment by assuring him: (1) that retrenchment, as in his case, often falls on the best men; (2) that he could do much more for the tea industry as an independent worker with adequate scope than as a member of the obsolete organization he had just left; and (3) that a promising line of future work lay in the systematic conversion into humus of the waste products of the tea estates. He agreed. Then Providence intervened on his behalf, on behalf of the tea plant and of the tea industry. Dr. Harler was offered and accepted (August 1933) the post of Scientific Officer to the Kanan Devan Hills Produce Company in the High Range, Travancore, the property of Messrs. James Finlay & Co. Ltd., who direct the largest group of tea gardens in the world. On taking up his duties at Nullatanni near Munnar, Dr. Harler proceeded to apply the Indore Process on an estate scale. No difficulties were met with in working the method;

ample supplies of vegetable wastes and cattle manure were available; the local labour took to the work and soon the General Manager of the Company, as well as the Estate Managers, became enthusiastic. It was now possible to pose the following question to the tea plant: What do you need to throw off disease and to do your best as regards the yield and quality of tea?

The second half of this question was soon answered on the Kanan Devan tea gardens, the first half had to wait till some years later. The pioneering work at Nullatanni, which was completed towards the end of 1934, was followed by the adoption of the Indore Process on the rest of the gardens—some forty in number. Each garden made from its available vegetable and animal wastes all the manure the tea needed; no artificials were necessary; yield and quality notably improved. But the tea plant in these gardens could say nothing about its requirements to ward off disease for the simple reason that with one small exception— the minor root trouble referred to below—there was practically no disease to resist in these well managed properties. All that properly made compost could do was to increase the yield and improve the quality of the tea above the high standard already reached.

When the news of Dr. Harler's successful estate-scale trial at Nullatanni reached me in September 1934, it occurred to me that it might be worth while bringing the possibilities of the Indore Process to the notice of the rest of the tea industry, which is arranged in large groups controlled by a small London directorate principally recruited from the industry itself. As I had no contacts with these bodies it was necessary to make one—preferably with some pioneer likely to be interested. I soon found the man—Mr. James Insch, one of the then Managing Directors of Messrs. Walter Duncan & Company. A small-scale trial of the Indore Process was completed on fifty-three estates in this group in Sylhet, Cachar, the Assam Valley, the Dooars, Terai, and the Darjeeling District. By the beginning of 1935 some 2,000 tons of compost in all were made and distributed. Five years later the quantity on the Duncan group had passed the 150,000 tons a year mark. But again the tea plant on these widely distributed properties did not answer the question: What do you need to throw off disease? The reason for this was that, as on the High Range of Travancore, the amount of disease on these estates was insufficient for such a question to be posed and answered. On these properties all the Indore Process could do was to raise the yield and improve the quality still further.

113

The results already referred to and the publicity they received came to the notice of many other groups of tea estates in India, Ceylon, and Africa. The methods of composting which had proved so successful on the Finlay and Duncan estates were tried at many new centres. It was in the course of these widely dispersed trials that the tea plant informed us what it needed to keep insect and fungous pests in check and why it wanted this assistance.

In a few cases during this third series of trials both insect and fungous diseases did occur to an extent which reduced somewhat the yield of tea. There was just sufficient disease here and there for the query under discussion to be put to the tea plant. The question on these particular gardens was not posed deliberately, but quite by accident. While this series of trials was in progress, example after example came to my notice in which such small applications of compost as five tons to the acre were at once followed by a marked improvement in growth, in general vigour, and in resistance to disease. Although very gratifying in one sense, these results were distinctly disconcerting. If humus acts only indirectly by increasing the fertility of the soil, time will be needed for the various biological, physical, and chemical changes to take place. If the plant responds at once, as was obviously the case, some other factor besides a general improvement in soil fertility must be at work. What could this factor be? It was clearly some agency which enabled humus to effect directly and very quickly the nutrition of the plant.

In a circular letter issued on 7th October 1937 to correspondents in the tea industry I suggested that the most obvious explanation of any sudden improvement in tea observed after one moderate application of compost could only be due to the effect of humus in stimulating the mycorrhizal relationship, which I afterwards discovered had been observed in Java in the roots of this crop. It seemed to me that this association must be present and that it would enable the fungous factor in the partnership to transfer the digestion products of protein into the sap and then into the green leaf. The virtues of humus could thus be moved from soil to plant in a very short space of time. This would enable the plant not only to resist disease, but would also explain the marked improvement in the yield and quality of tea which resulted from dressings of compost. I saw all this in imagination, as it were, on 7th October 1937 as a likely hypothesis to explain the facts. What set these ideas in train was a perusal of Dr. M. C. Rayner's work on conifers at Wareham [1] in Dorsetshire, where small additions of properly made compost has led to spectacular results most easily explained by the establishment of the mycorrhizal association.

At this juncture a group of tea companies which had adopted the Indore Process asked me to visit their estates in India and Ceylon. In the course of this tour, which lasted from November 1937 to February 1938, I examined the root system of a number of tea plants which had been manured with properly made compost, and found everywhere the same thing—numerous tufts of healthy-looking roots associated with rapidly developing foliage and twigs much above the average. Both below and above ground humus was clearly leading to a marked condition of well-being. When the characteristic tufts of young surface roots were examined microscopically, the cortical cells were seen to be literally overrun with mycelium to a much greater extent than is the rule in a really serious infection by a parasitic fungus. Clearly the mycorrhizal relationship was very much involved: my hypothesis was abundantly confirmed: the tea plant had a message to deliver on the disease question. My hasty and imperfect observations made in the field and in the course of a very strenuous tour—during which many estates were visited in detail and many lectures were delivered to groups of planters—were confirmed and extended by Dr. M. C. Rayner and Dr. Ida Levisohn who examined a large number of my root samples, including a few in which artificials only were used or where the soils were completely exhausted and the garden had become derelict with perhaps only half the full complement of tea plants. In these latter cases the characteristic tufts of normal roots were not observed; development and growth were both defective; the mycorrhizal association was either absent or poorly developed. Where artificials were used on worn-out tea, infection by brownish hyphae of a *Rhizoctonia*-like fungus (often associated with mild parasitism) was noticed. But whenever the roots of tea manured with properly made compost were critically examined, the whole of the cortical tissues of the young roots always showed abundant endotrophic mycorrhizal invasion, the mainly intra-cellular mycelium apparently belonging to one fungus. Thus fungus was always confined to the young roots and no invasion of old roots was observed. In the invaded cells the mycelium exhibits a regular cycle of changes from invasion to the clumping of the hyphae around the cell nuclei, digestion and dis-

[1] An account of this Wareham work has since been published in 1944 in book form under the title—*Problems in Tree Nutrition*—by Messrs. Faber and Faber, London.

integration of their granular contents, and the final disappearance of the products from the cells. In this way the digestion products of the proteins of the fungus pass into the cell sap and then into the green leaves.

Humus in the soil, therefore, affects the tea plant direct by means of a middleman—the mycorrhizal association. Nature has provided an interesting piece of living machinery for joining up a fertile tea soil with the plant. Obviously we must see that this machinery is provided with the fuel it needs—continuous dressings of properly made compost. I saw on several occasions the response of the tea plant, which had been attacked by disease, to small dressings of compost. I was amazed by the way even a single application had reduced the amount of infection and started the tea bushes well on the way to complete recovery.

The tea plant had now answered the question: What must be done to me to be saved? It is nothing less than the restitution of the manurial rights this plant enjoyed in its forest home—regular supplies of freshly prepared compost.

One difficulty was encountered and partly overcome in this restitution of manurial rights. In some of the tea areas the gardens were so closely jammed together that it was not possible to maintain the head of cattle needed to provide the animal manure for making first-class compost. I suggested that in such cases pigs would be the easiest livestock to keep and that the cost of the pig food brought on to the gardens could be found by reducing the amount of artificial manure that would be needed. But where land was available, steps were taken to increase the head of other livestock to make the necessary animal manure.

One interesting case of introducing cattle into the tea gardens solely for their manure came to my notice from Africa. When Viscount Bledisloe returned to England from his African mission, where he had been Chairman of a Royal Commission connected with the affairs of the Rhodesias and Nyasaland, he presented me with an enlarged set of the photographs he had taken on compost making, the virtues of which he constantly brought to the notice of the various local governments with whom he came in contact. In this way he did much of the spade work which was necessary to make South Africa compost-minded. One of these photographs, taken at Messrs. J. J. Lyons & Company's estate at Mlange, showed the cattle which the tea gardens of Nyasaland were beginning to keep solely for compost making (Plate III). This, indeed, was proof positive of progress and of enterprise. If the tea gardens of Africa

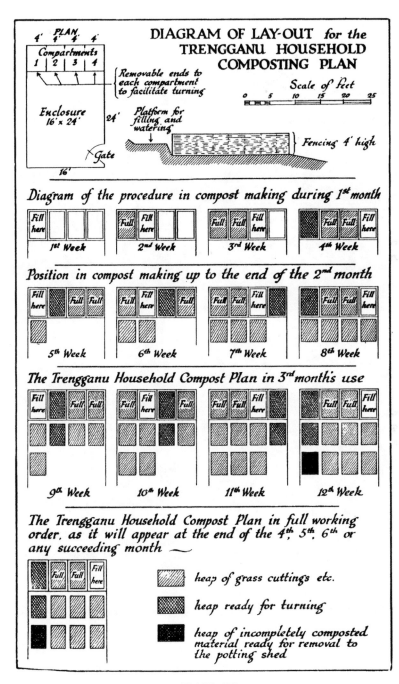

PLATE VII

COVERED AND UNCOVERED PITS

ROOFING A PIT

PLATE VIII. COMPOSTING AT GANDRAPARA

can go to the trouble of maintaining cattle for the sake of the urine and dung they produce, what is to prevent other plantation industries all over the world doing the same? It is impossible to farm for long without livestock. It is equally impossible to maintain the overseas plantations in an efficient condition without these living manure factories for producing two of the essentials for making humus. Like tea, all these plantation crops—coffee, cacao, sugar-cane, cotton, sisal, maize, coconuts, bananas, citrus fruit, grapes, apples, pears, peaches, and so forth—are mycorrhiza formers. All need the digestion products of fungous protein to maintain the power to reproduce themselves, to provide high-quality crops, and to resist the onslaught of insects and fungi.

But cases of disease occur in tea which cannot be remedied by getting the surface soil into good fettle. The tea is a deep-rooting plant and makes great use of the lower roots to keep up the water supply during dry weather. These deep roots must, therefore, function properly. There must be no waterlogging due to stagnant water held up by impermeable layers in the subsoil. This condition invariably results in root disease duly followed by the death of the plant. The only example of such disease of any consequence I met with during my second tour in India and Ceylon was a root fungus which appeared here and there and destroyed the bushes over small areas particularly on the laterite soils of South India. The real cause of the trouble appeared to be some interference with drainage in the lower layers of the soil, which reduced the vitality of the tea and prepared the way for the parasite. Such diseases might be dealt with most easily by Swedish pillar-drains—vertical pits, dug well below the layer under the laterite holding up the stagnant water, and afterwards filled with large stones.

At the Gandrapara estate on the flat stretches of the alluvium of the Bengal Dooars I saw one of the best examples in my experience of successful surface drainage under a high monsoon rainfall, which I was told had proved very useful in the prevention of root disease. On this fine property, very deep and narrow minor earth drains had been constructed among the tea and connected up with wider major ditches which carried off the surplus water to the natural drainage lines. The system was based on a contour survey and had been carried out by a competent engineer. The minor drains could not easily be detected, as the tea bushes on either side met above the drains, forming everywhere a continuous green table. With the combined help of the excellent top shade and this green table the heavy monsoon downfalls were converted into fine spray, which was readily absorbed by the heavily composted surface soil without any great silting up of these minor drains. I had studied surface drainage in many parts of the world, including some of the best examples Italy has to provide, and had carried out drainage schemes on the land in my own charge, but none of these came up to the Gandrapara standard. I mentioned this fact at a lecture to a group of local tea planters at Gandrapara. By chance the engineer who had designed the local scheme was present. His grateful reaction to my chance remarks will remain as one of my pleasantest recollections.

The superficial character of the conventional investigations on the diseases of tea will be clear from what has been set out above. Nothing is to be gained by starting research on any future tea disease at the wrong end. Investigation must always begin with the soil. If the mycorrhizal association is not working properly, this must be put right in the first place. The drainage of the soil round the deep roots must also be effective. In all probability the result will be rapid disappearance of pests. Proceeding in this way, diseases can be made very useful for keeping a tea garden up to the mark as regards manuring and soil management.

CACAO (*THEOBROMA CACAO*)

A good deal of time was spent by me in Grenada about 1901 on the study of the fungous diseases of cacao. Visits were also paid to a number of cacao estates in Trinidad and Dominica. The main troubles were three: die-back of the leaders on low-lying areas (caused by poor drainage), pod, and bark diseases. A new fungous pest—the witch broom disease—had just made its appearance in Surinam, but had not then spread to Trinidad and the other islands. It has since become a serious trouble in the West Indies.

Among the many estates visited was a small plantation in Grenada owned by the late Rev. G. W. Branch, which stood out from the rest of the island by virtue of the heavy yields of high-quality beans; the fact was ascertained that these cacao trees were always manured with farmyard manure. Although a paper was read by the owner at one of the West Indian Conferences in the early years of this century and full details of the method of manuring were given, it never struck anyone that here in a nutshell was the solution of the main problem of cacao, namely, mixed farming and the preparation of plenty of freshly prepared compost for the cacao trees. Everybody without exception who attended this meeting was labouring under the thraldom of the NPK mentality and was only able to think in terms of so many pounds to the acre of this or that artificial manure. Though many were impressed by these Grenada results, they seemed incapable of facing up to their very obvious implications. All this happened about 1901.

In 1908 in the course of a visit to Ceylon I saw these Grenada results repeated, but on a much larger scale, at the Kondesalle cacao estate near Kandy. Thirty years later—in 1938—when on my tour of the tea estates of India and Ceylon I resumed my interest in cacao and re-visited Kondesalle, at which the finest cacao beans I have ever seen are being produced. I again observed no cacao diseases on this property and was not told of any by the manager or by his assistants. The trees appeared exceeding healthy and here again, as on the small Grenada plantation, livestock—in this case, pigs and Hissar cattle—were kept for producing the farmyard manure applied to the cacao trees.

During this tour samples of the surface roots of cacao at Kondesalle were fixed and sent to London for examination by Dr. Rayner. The results are referred to in my report on this tour in the following words:

"*Cacao*. Dr. Rayner examined the surface roots of cacao from Kondesalle (Ceylon) taken from a field which had been manured with farmyard manure. Sporadic mycorrhizal infection of endotrophic (i.e. intracellular) type was present. Compost is not yet being made on this estate. It will be interesting to see whether still better results than those now yielded by farmyard manure on this fine property could not be obtained if the cattle and pig manure were first composted with the estate wastes and used in the form of humus."

It will be obvious that in both Grenada and Ceylon examples of how to grow heavy crops of high quality cacao, free from disease, have long been provided by accident, as it were. Meanwhile both these regions have been furnished with modern agricultural departments. The astounding fact is that no one in these organizations or in the planting community has understood the value or the significance of the lessons these two estates have to teach. Nevertheless, both indicate quite clearly how cacao will have to be produced in the future if the growing menace of disease is to be averted. As is well known, much of the cacao of commerce now comes from West Africa, where it is produced largely at the expense of the original stores of humus left by the forest. As in Grenada and Trinidad, these stores will not last for ever. After a time they will be used up and the day of reckoning will arrive. Indeed, this has already come.

In the *West India Committee Circular* of September 1944, an article appeared on the future welfare of this crop in the Gold Coast—the world's largest exporter of cacao. It appears that the industry is face to face with a crisis "perhaps without equal in the history of any major tropical crop in the British Empire."

Two factors are responsible for this state of affairs: (1) the swollen-shoot virus disease, first reported in 1936, and (2) capsid bugs. These two pests are being investigated at the Tafo Cacao Research Station established by the local Agricultural Department in 1938. The spread of these two diseases has been so rapid as to constitute a direct menace to the whole future of the industry. In 1943 a conference of research workers was held at Tafo, presided over by the Agricultural Adviser to the Secretary of State. A programme of future research in cacao was formulated. Plans were also made for the reorganization of the Tafo Station as the West African Cacao Research Institute, for which a director has been appointed.

There seems no doubt that what is needed to place the cacao industry of the Gold Coast on a sound foundation is not more research into cacao diseases, but the introduction of livestock into the areas growing cacao and the conversion of the wastes of the animal and the plant into humus, as Messrs. J. J. Lyons & Company have done on their tea estates in Nyasaland (p. 116). The Gold Coast cacao industry, which began to export produce at the beginning of the century, has obviously been living for the last forty years or so on capital—on the humus left by the original forest. This has now been used up and Nature has registered her usual protest in the form of disease. The West African cacao trees have been deprived of their manurial rights. The Kondesalle cacao estate in Ceylon indicates what should be done to put matters right. No committees, however well selected, and no amount of research, however devoted, will alter this obvious conclusion. The time has indeed come for the prodigal to return, to confess, and to start proper farming.

There is no doubt that the cacao industry all over the Empire could at once be restored by mixed farming and the systematic conversion into compost of all the vegetable and animal wastes available. The manufacturing interests in Great Britain which need a regular and reliable supply of cacao beans should at once use their influence and insist that this obvious reform be taken in hand forthwith.

One objection to this suggestion must be answered in advance. If a portion of the existing areas under cacao is devoted to mixed farming, how is the output to be maintained? The answer is: By virtue of the vastly increased yield and better quality of the beans, as well as the longer life of the trees. There is ample land in all the cacao-growing areas of the Empire for this crop and also for livestock: there is no reason why this reform should not be set in motion forthwith. Must we always wait for catastrophe before the simplest step forward can be taken? What has the agricultural research organization of the Colonies been doing to allow such a state of affairs as this Gold Coast cacao scandal to develop?

COTTON

The cotton crop suffers from many insect and a few fungous diseases. It has already been mentioned that one-quarter of the space of the last pre-war issues of the *Empire Growing Cotton Review* was devoted to disease. The alarming significance of the figures given can only be realized when it is remembered that cotton is a distinctly robust crop that does not need very intensive methods of farming to produce fair yields of fibre. Moreover, cotton should not exhaust the land very much, as the fibre of commerce contains little more than the cellulose manufactured from the gases of the atmosphere and the water in the soil; the flowers fall after the bolls set; the leaves of the crop mostly drop before the stalks are removed; the roots remain in the ground: the seed is very useful for feeding the work cattle. Provided, therefore, a fair proportion of the cotton seed is passed through the stomachs of oxen and other animals and the old stalks find their way back to the soil in the form of humus, this crop cannot possibly wear out the land to any appreciable extent. Further, as inter-cultivation between the rows has to stop when the flowers appear, a cotton crop always enables weeds to cover the surface which, when ploughed under, help to maintain the humus content of the soil. If the incidence of disease depends on the poverty of the soil, it would seem that there must be something very wrong somewhere in the current methods of cotton growing; otherwise these diseases ought not to occur. A cotton crop, if properly looked after, ought to be very free from pests.

During the years 1924–31 I had unique opportunities for the study of this crop, because during this period I held the post of Director of the Institute of Plant Industry at Indore in Central India, at which cotton was the principal crop. Indeed, the new institute could not have been founded or maintained without the help of large grants from the Indian Central Cotton Committee, which in turn was financed by a small annual cess on each bale of raw cotton exported from India or used in the local mills. This cess was naturally passed on to the multitude of small-holders who raised the crop. If, therefore, the Indian Central Committee could do something to help these men in return for their money, this new body and its various research workers would have justified their existence.

Before taking up an investigation of the cotton crop at Indore in 1924, a survey of cotton growing in the various parts of India was undertaken. At the same time, the research work in progress on cotton in other parts of the world was critically examined.

As regards cotton growing in India, the two most important areas are: (1) the black cotton soils of the Peninsula, which are derived from the basalt; (2) the alluvium of north-west India, consisting of deposits left in a deep chasm by the rivers of the Indo-Gangetic plain. Besides these there are small areas of garden cultivation in southern India, where American types of cotton are grown intensively under irrigation and where heavy crops of good fibre are the rule.

On the black soils there are thousands of examples which indicate the direction research on this crop should take. All round the villages of the Peninsula, zones of very highly manured land, rich in organic matter, occur. These are kept in good fettle by the habits of the people: the night-soil is habitually added a little at a time to the surface of the fields. On such zones cotton does well no matter the season; the plants are well grown and remarkably free from pests; the yield of seed cotton is high. On the similar but unmanured land alongside the growth is comparatively poor; only in years of well-distributed rainfall is the yield satisfactory. But even under the most adverse conditions one is amazed to see how the cotton plant manages to survive and to produce some kind of crop. Only the very hardiest plant could produce seed under such unfavourable circumstances. The limiting factor in growth on these black soils is the development, soon after the rains set in, of a colloidal condition, which interferes with aeration and impedes percolation. This occurs on all black soils, but organic matter mitigates the condition. As these soils dry out at the end of the rains, extensive cracking occurs which aerates the soil but also damages the roots and rapidly desiccates the soil. The varieties of cotton, therefore, must possess the power of rapid ripening, otherwise the bolls could not open in time. The growth period of any successful cotton on the rain-fed, black soil areas must be short; the plant must literally burst into cotton at picking time and show no tendency to linger in yielding up its crop. Two pickings at the most are all that is possible.

On the alluvium of north-west India a somewhat similar limiting factor occurs. Here cotton is grown on irrigation, which first causes the soil particles to pack and later on to form colloids. In due course the American varieties, whose root systems, compared with those of the indigenous cottons, are superficial, show by their growth that they are not quite at home. The anthers, the most sensitive portion of the flower, sometimes fail to open and to release their pollen: the crop is unable to set a full crop of seed. But this is not all. The ripening period, particularly in the Punjab, is unduly prolonged; as many as four pickings are necessary. Moreover, the fibre often lacks strength, quality, and life. The cause of these troubles is poor soil aeration, which in these soils leads to a very mild alkali condition. This, in turn, prevents the cotton crop from absorbing sufficient water from the soil. One of the easiest methods of preventing this packing and alkali formation is to increase the bacterial population by means of dressings of humus. In this way the soil is able to re-create a sufficient supply of compound particles to restore the aeration and improve the water supply needed by the cotton.

As regards disease, insects cause more damage to the crop than do fungi: there is more insect disease on the alluvium than on the black soils. The insect diseases on the alluvium mostly affect the bolls which, as we have seen, develop but slowly. If the cotton could be made to ripen more quickly, these boll diseases might be very considerably reduced.

The direction of research work on cotton was, therefore, disclosed by a study in the field of the crop itself. The problem was how best to maintain soil aeration and percolation. This could be solved if more humus could be obtained. At the same time, there appeared to be every chance that more humus would materially reduce, by speeding up maturation, the damage done to the ripening bolls by the various boll worms. Good farming methods, therefore, including a proper balance between livestock and cotton, seemed to provide the key to the cotton problems of

India. Once the soils were got into good fettle and maintained in this condition, the question of improved varieties could then be taken up with every chance of success. To hope to overcome bad farming by improving the variety in the first place was an obvious impossibility, such a research policy amounting to a contradiction in terms.

A study of the research work on cotton which had been done all over the world did nothing to modify this opening. Cotton investigation everywhere appeared to suffer from the fragmentation of the factors, from a consequent loss of direction, from failure to define the problems to be investigated, and from a scientific approach on far too narrow a front without that balance and stability provided by adequate, first-hand farming experience. The research workers seemed to be far too busy on the periphery of the subject and to be spending their time on unimportant details. This has naturally resulted in a spate of minor papers which lead nowhere except to the cemetery so providentially furnished by the *Empire Cotton Growing Review*. In Africa, particularly, much time and money have been wasted in trying to overcome, by plant-breeding methods, diseases which obviously owe their origin to a combination of worn-out soil and bad farming.

Steps were therefore taken at Indore to accelerate the work on the manufacture of humus which had been begun at the Pusa Research Institute. The Indore Process was the result. It was first necessary to try it out on the cotton crop. The results are summed up in the following table.

THE INCREASE IN GENERAL FERTILITY AT INDORE

Year	Area in acres of improved land under cotton	Average yield in lb. per acre	Yield of the best plot of the year in lb. per acre	Rainfall in inches
1927	20.60	340	384	27.79 (distribution good)
1928	6.64	510	515	40.98 (a year of excessive rainfall)
1929	36.98	578	752	23.11 (distribution poor)

The figures show that, no matter what the amount and distribution of rainfall were, the application of humus soon trebled the average yield of seed cotton—200 lb. per acre—obtained by the cultivators on similar land in the neighbourhood.

In preparing humus at Indore one of the chief wastes was the old stalks of cotton. Before these could be composted they had to be broken up. This was accomplished by laying them on the estate roads, where they were soon reduced by the traffic to a suitable condition for use as

124

bedding for the work cattle prior to fermentation in the compost pits. I owe this suggestion to Sir Edward Hearle Cole, who hit upon this simple device on his Punjab estate.

The first cotton grower to apply the Indore Process was Colonel (now Sir Edward) Hearle Cole at the Coleyana Estate in the Montgomery District of the Punjab, where a compost factory on the lines of the one at the Institute of Plant Industry at Indore was established in June 1932. At this centre all available wastes have been regularly composted since the beginning; the output is now about 8,000 tons of finished humus a year. Compost has increased the yield of cotton, improved the fibre, lessened disease, and reduced the amount of irrigation water by a third. The neighbouring estates have all adopted composting; many interested visitors have seen the work in progress. One advantage to the Punjab of this work has, however, escaped attention, namely the importance of the large quantities of well grown seed, raised on fertile soil, contributed by these estates to the seed distribution schemes of the Provincial Agricultural Department. Plant breeding, to be successful, involves two things—an improved variety plus seed for distribution grown on soil rich in humus.

The first member of an agricultural department to adopt the Indore method of composting for cotton was Mr. W. J. Jenkins, C.I.E., when Chief Agricultural Officer in Sind, who proved that humus is of the greatest value in keeping the alkali condition in check, in maintaining the health of the cotton plant, and in increasing the yield of fibre. At Sakrand, for example, no less than 1,250 cart-loads of finished humus were prepared in 1934–5 from waste materials such as cotton stalks and crop residues.

During recent years the Indore Process has been tried out on some of the cotton farms in Africa belonging to the Empire Cotton Growing Corporation. In Rhodesia, for example, interesting results have been obtained by Mr. J. E. Peat at Gatooma. These were published in the *Rhodesia Herald* of 17th August, 1939. Compost markedly improved the fibre and increased the yield not only of cotton, but also of the rotational crop of maize. The results obtained by the pioneers in India, therefore, apply to Africa.

Why cotton reacts so markedly to humus has only recently been discovered. The story is an interesting one, which must be placed on record. In July 1938, I published a paper in the *Empire Cotton Growing Review* (Vol. XV, No. 3, 1938, p. 186), in which the role of the mycor-

125

rhizal relationship in the transmission of disease resistance from a fertile soil to the plant was discussed. In the last paragraph of this paper the suggestion was made that mycorrhiza "is almost certain to prove of importance to cotton and the great differences observed in Cambodia cotton in India in yield as well as in the length of the fibre, when grown on (1) garden land (rich in humus) and (2) ordinary unmanured land, might well be explained by this factor." In the following number of this *Journal* (Vol. XV, No. 4, 1938, p. 310) I put forward evidence which proved that cotton is a mycorrhiza former. The significance of this factor to the cotton industry was emphasized in the following words:

"As regards cotton production, experience in other crops, whose roots show the mycorrhizal relationship, points very clearly to what will be necessary. More attention will have to be paid to the well tried methods of good farming and to the restoration of soil fertility by means of humus prepared from vegetable and animal wastes. An equilibrium between the soil, the plant, and the animal can then be established and maintained. *On any particular area under cotton, a fairly definite ratio between the number of livestock and the acreage of cotton will be essential.* Once this is secured there will be a marked improvement in the yield, in the quality of the fibre, and in the general health of the crop. All this is necessary, if the mycorrhizal relationship is to act and if Nature's channels of sustenance between the soil and the plant are to function. Any attempt to side-track this mechanism is certain to fail.

"The research work on cotton of to-morrow will have to start from a new base line—soil fertility. In the transition between the research of to-day and that of the future, a number of problems now under investigation will either disappear altogether or take on an entirely new complexion. A fertile soil will enable the plant to carry out the synthesis of proteins in the green leaf to perfection. In consequence the toll now taken by fungous, insect, and other diseases will at first shrink in volume and then be reduced to its normal insignificance. We shall also hear less about soil erosion in places like Nyasaland, where cotton is grown, because a fertile soil will be able to drink in the rainfall and so prevent this trouble at the source."

Confirmation of these pioneering results soon followed. In the *Transactions of the British Mycological Society* (Vol. XXII, 1939, p. 274) Butler mentions the occurrence of mycorrhiza as luxuriantly developed in cotton from the Sudan and also in cotton from the black soils of

Gujerat (India). In the issue of *Nature* of 1st July 1939, Younis Sabet recorded the mycorrhizal relationship in Egypt. In the *Empire Cotton Growing Review* of July 1939, Dr. Rayner confirmed the existence of mycorrhiza in samples of the roots of both Cambodia and Malvi cotton collected at my suggestion for her by Mr. Y. D. Wad at Indore, Central India, from both black cotton soil and from sandy soil from Rajputana.

The problem now to be solved in cotton production and in the control of disease is the discovery of the easiest way in which the present extensive methods of agriculture can be converted into more intensive methods. This involves a great increase in livestock in the existing cotton areas and the systematic conversion of the cotton stalks into humus. In this way the yield per acre can rapidly be increased and the fibre improved. The present supplies of cotton can, therefore, be produced from about two-thirds the area now under this crop. The land so released can be used for the production of food grains and fodder crops. A balanced agriculture is the key to the prevention of the diseases of cotton.

Every point here discussed was mentioned or suggested in the section on cotton in *An Agricultural Testament* published in 1940. It will be interesting to observe how long it will take such bodies as the Empire Cotton Growing Corporation and the Indian Central Cotton Committee to revise their research policies and to replace their laboratory workers by farmer-scientists.

RICE

The most important cereal in the world is rice. Moreover, it is a crop remarkably free from diseases of all kinds. Rice, therefore, should take high rank among Nature's professors of agriculture. A study of its cultivation might teach us much about the prevention of disease.

But the moment we embark on such a study we find no less than three of the principles underlying Western agricultural science flatly contradicted by this ancient cultivation.

In the first place, in many of the great rice areas of the world there is no such thing as a rotation of crops. Rice follows rice year after year and century after century without a break, without even a fallow year every now and then. Moreover, there is no falling off in yield and no sign of soil exhaustion. There is, therefore, no need of a continuous rice experiment of the Broadbalk pattern for the simple reason that such age-long experiments are to be seen everywhere. To begin a new one would be to carry coals to Newcastle.

In the second place, these continuous rice crops do not need those extraneous annual applications of nitrogenous manures which are considered to be essential for all cereals. The rice fields somehow manure themselves.

In the third place, the rice crop often covers vast areas of land in one unbroken sheet, thereby providing a paradise for insect and fungous diseases. But these do not occur: on the contrary, the rice crop is generally remarkably free from diseases of all kinds.

What is the secret underlying these unexpected and unconventional results? The beginning of the solution of the riddle will, I think, be found in the nurseries in which the young rice plants are raised before transplanting. These are always on well aerated and well manured land, the manure, as a rule, being well decayed cattle manure. The result is the rice seedlings become veritable arsenals of such things as nitrogen, phosphorus, and potash, all in organic combination. Moreover, the rice plant is a mycorrhiza former and so ample provision occurs even in the seedling stage for the circulation of protein between soil, sap, and green leaf. How important this building up of the rice seedling is will be clear, when it is realized that the transplanting process from well aerated soil to mud involves a completely fresh start in a new environment. This results in a delay of many days and, therefore, in the loss of a substantial proportion of the total growing period. Nevertheless, transplanting pays, because transplanted rice always gives a better yield than broadcast rice in which, of course, there is no delay in growth. Here we have a clear and definite lesson from the long experience of the Orient, namely, the vital importance of well-nourished seedlings. This applies in particular to crops like fruit, tea, coffee, cacao, tobacco, vegetables, and so forth. In all these well begun is half done.

But how does the rice manage to manure itself? The answer is provided by the nitrogen-fixing powers of the algal film found in rice fields. This algal film does three things: it aerates the water of the rice fields; it fixes a continuous supply of nitrogen from the atmosphere; it leaves behind a useful amount of easily decomposable organic matter. Nevertheless, more organic matter is needed in the rice fields beyond that supplied by the algal film and the roots of the old crop. How markedly rice benefits from compost has been proved at Dichpali in India. The results have already been set out in Chapter V of *An Agricultural Testament*, pp. 80–2.

The problem now is to find more compost for the rice crop. Nature

128

has already provided ample vegetable waste in the shape of the water hyacinth, an aquatic weed to be found in most of the rice-growing areas of the world. This water weed should be regarded as a heaven-sent gift of Providence for the rice-growing areas, as it provides not only large supplies of readily fermentable vegetable matter, but sufficient moisture for the composting process as well. All that is needed besides is a supply of cow-dung and urine earth, both of which are available locally. In Bengal, for example, the annual yield of rice could be vastly increased if only a national campaign for the composting of the water hyacinth could be set in motion. That this weed makes excellent compost has already been fully demonstrated: first at Barrackpore, near Calcutta, by Mr. E. F. Watson, O.B.E., the Superintendent of the Governor's Estates, Bengal, and later on some of the tea estates in Assam. No future rice famines in Bengal need be feared once full use is made of the vast local supplies of water hyacinth.

What is the explanation of the comparative immunity of the rice crop from disease? I think the answer is provided by the fact that rice is a mycorrhiza former and that this mechanism works not only in the rice nurseries, but also in the paddy fields themselves: nothing has interfered with this process, as artificial manures are unknown and such bad practices as over-irrigation are, from the nature of the case, impossible. Indeed, the behaviour of this crop as regards parasites supplies strong confirmation of the view that what matters most in crop production is the effective circulation of protein between soil and sap, followed by the synthesis of still more protein of the right kind in the green leaf. *High quality protein will, in ordinary circumstances, always protect the plant against its enemies.*

WHEAT

For nineteen years, 1905–23, I was engaged in a study of the wheat crop of India, which included work on the creation of new varieties. The records of the work on Indian wheat carried out at Pusa will be found in *Wheat in India,* published in 1908, and in a series of thirty-four papers issued by the Agricultural Research Institute, Pusa. A list of these papers will be found in *The Application of Science to Crop Production,* Oxford University Press, 1929, and a summary in *Bulletin 171* of the Agricultural Research Institute, Pusa, 1928.

Pusa is situated near the eastern extremity of the area under this crop, where the wheat and rice tracts are intermingled and where there is more rice than wheat. As would be expected, both the soil and atmospheric conditions are distinctly on the damp side for wheat. All three of the common rust fungi—brown, yellow, and black rust—were much in evidence. In one respect this was an advantage in plant breeding. It was easy to arrange for abundant infecting material for testing the reaction of the various cultures to these parasites. I did nothing to destroy these rusts; I did everything possible to have them always at hand. The result was that my ideas as to the cause of fungous diseases were constantly being verified. If a variety of wheat is resistant to one or more of these rusts, it makes no difference at all how much infecting material rains upon it or how much diseased stubble is ploughed into the land. Nothing happens even in wet seasons which always favour infection.

In the course of this work some interesting observations on immunity were made. Among the types of wheat in the submontane tracts of North Bihar a number were found which were very seldom or never attacked by rust. They were, to all intents and purposes, immune. Unfortunately they all possessed weak straw and poor yielding power, and were only useful as plant breeding material. Should, in the future, any wheat breeder need such types, they could either be collected at harvest time or selected from the crop raised from bazaar samples of wheat from this tract.

Another wheat which was immune to all three rusts was the primitive species known as *einkorn (Triticum monococcum)*. But this wheat never flowered at Pusa, remaining in the vegetative condition till harvest time. One year some of these dense tufts were allowed to remain in the ground till the rains broke in June. This species was not killed by the intense hot weather of April and May, but as the hot season developed it began to show signs of infection by some parasite. This proved to be black rust—an interesting example of the destruction of immunity by adverse weather conditions, and a very striking confirmation of Mr. J. E. R. McDonagh's views on the limits of immunity set by extreme climatic conditions (p. 187).

The most interesting case of wheat disease I met with in my tours was in an area of low-lying land in the Harnai valley in the mountains of the Western Frontier. Here I found wheat growing in wet soil, in which the aeration was poor and the general soil conditions more suitable for rice than for wheat. It appeared this area was always affected by eelworm, which, however, never spread to the adjoining wheat areas which continued almost without a break for at least 1,000 miles to the east. Through this valley there was a constant stream of all kinds of traffic both ways—towards Afghanistan to the west and towards the great cities of the plains in the east. Nothing was done to check the infection of the neighbouring wheat areas by preventing the cysts of the eelworm being carried by the feet of animals or men or by wheeled traffic. Infection both ways must have been going on without interruption for hundreds of years. But nothing had happened. Obviously the eelworm is not the cause of the trouble or no power on earth could have stopped the whole of the wheat areas of a sub-continent becoming infected. Before infection is possible the soil conditions must be favourable.

A similar case of eelworm on rice occurred in the deep-water rice areas of Bengal, where the disease is known as *ufra*. Again we have a heavily infected area in close contact with one of the greatest rice areas of the world. No precautions are taken to isolate the area and protect the surrounding rice from infection. There has been no spread of the trouble outside the small deep-water areas which favour the eelworm.

These two outstanding cases, I think, dispose of the eelworm bogey, which threatens to raise its head in Great Britain in connection with the eelworm diseases of potato and sugar beet. The experts propose measures to control the potato crop so as to prohibit the movement of tubers from and into certain areas. They also recommend that infested areas should give up growing these crops for some years till the eelworm dies out naturally. Before these suggestions are accepted by the authorities consideration might be given to the significance of the two cases—wheat and rice—cited above, and also to the elimination of eelworm on farms and gardens in Southern Rhodesia by dressings of freshly prepared compost (p. 153).

Intimately bound up with the resistance of the growing wheat plant to disease is the way wheat straw can stand up to the processes of decay when used as thatch. Is there any connection between the life of a thatched roof and the manurial treatment of the land which produced the wheat straw? There is. Farmyard manure results in good thatch, artificials in bad thatch. This will be evident from the following extracts from an article entitled "Artificial Manures Destroy Quality," which appeared in the *News-Letter on Compost,* No. 4, October 1942, p. 30:

"In the case of the wheat crop raised on Viscount Lymington's estate in Hampshire, careful records have been kept of the life of wheat straw when used for thatching. Wheat straw from fields manured with organic matter, partly of animal origin, lasts ten years as thatch; straw from similar land manured with artificials lasts five years."

Interesting confirmation of this view on the life of wheat straw in thatch has been supplied in a recent letter dated 10th September 1942, from a correspondent (Mr. J. G. D. Hamilton, Jordans, Buckinghamshire), who writes:

"About five years ago, while visiting craftsmen in Wiltshire, I was told by two old thatchers in different parts of the county that the straw they had to work with now was not nearly so good as that which they had had in years gone by. Both gave as the reason the modern use of artificials in place of farmyard manure."

Anyone owning a thatched building, who wishes to compare the virtues of compost with the harm done by chemical manures, can easily make use of the above experiences when the time comes to renew the roof. Alternate strips of the two kinds of straw will soon show interesting differences and will suggest a further trial—a comparison of the whole wheat bread made from the two samples of wheat.

VINE

One of the oldest crops in the world is the vine. Its original home is said to be in Central Asia whence it has spread everywhere. Even when outdoor conditions have made its cultivation impossible, it has been successfully grown under glass often, as in Holland, on a commercial scale. Such an ancient branch of crop production might, therefore, have much to teach us about disease and its prevention.

During some thirty years, from 1910 to 1939, I came in close contact with this crop, which I soon began to regard as one of my ablest teachers. The instruction I received falls naturally into three independent courses which can best be dealt with in order.

From 1910 to 1918, the summers of which were spent in the Quetta valley on the Western Frontier of India, I saw a good deal of grape growing in desert areas, as it had been successfully practised for many centuries. The tribesmen of Baluchistan select the well-drained slopes of the valleys for their vineyards, where the subsoil is sufficiently well aerated for healthy root development. The vines are grown in deep, narrow trenches, the excavated soil being piled on the undisturbed surface between to form ridges a few feet high, which break the force of the dry, hot winds which often sweep down these valleys. The floors of these trenches are well manured with farmyard manure, irrigated by flow when the vines are planted, after which they are supported by the steep earthen walls of the ditches. As the natural rainfall during the growth period is almost nil and as the trenches are naturally well drained, there is no danger of waterlogging. The amount of irrigation water needed is not excessive, as the trench system checks evaporation. The annual rainfall is mostly received in the form of snow, so that watering does not begin till after the buds break in the spring. These partly buried vineyards are invisible at a distance, as the vines are never allowed to grow above the ground level.

At first sight all the conditions necessary for fungous and insect diseases seemed to have been provided—a damp atmosphere round the vines and restricted air movement in the trenches. Nevertheless, there was no disease of any kind—at least I never found even the beginnings of such trouble. On the contrary, both the foliage and the wood exhibited every sign of robust health and well-being. The yield of grapes was heavy, the quality and keeping power excellent. Moreover, the varieties grown had been in cultivation for centuries. Nowhere did I hear of the activities of plant breeders in producing new types: no cases of the introduction of varieties from areas outside Central Asia came to my notice. Another characteristic of this cultivation, the significance of which was not fully appreciated till later, was never to cover the whole of the available area with vineyards. The tribesmen seemed to be content with a modest fraction of their land under grapes, leaving the remainder unused or devoted to crops like wheat. This enabled them to go in for mixed farming and to produce sufficient farmyard manure for their vines and other fruit. I saw no areas like many of the vine-growing regions of Europe, where every square foot of suitable land is devoted to grapes, leaving none to produce muck.

Under this system of cultivation the vine obviously flourished under semi-desert conditions; the crop possessed ample powers of disease resistance; the varieties to all intents and purposes were eternal; the fungicides, insecticides, spraying machines, and artificial manures of the West were unknown.

There was, however, one problem which needed investigation in Baluchistan. The grapes were not reaching the vast market provided by the cities of India, in spite of the fact that a direct broad-gauge railway line extended from the Afghan frontier at Chaman to all parts of the

133

sub-continent. This was due primarily to the primitive methods of packing in vogue. There was much waste of space in the railway fruit vans from the miscellaneous nature of the packages used, which were of all shapes, sizes, and weights. This naturally increased the freight rates. I was called upon to solve these problems and, although a Government official, obtained permission to trade in fruit so that I could discover at first hand the obstacles which had to be overcome. Two improvements were made: (1) the design and introduction of suitable crates, each containing twenty-four 2 lb. punnets of grapes, and (2) the unification of the rules of the many separate railway companies which handled the fruit, so that in view of the use of standard crates (by which the traffic could be easily handled and by which the revenue earned by each van could be increased) the empties were returnd free of charge. The non-returnable and returnable crates adopted for grapes and tomatoes, are illustrated in Fig. 3.

The problem then was to find the cheapest source of wood. This proved to be Norway. The Norwegian timber was cut up into suitable sections or made into punnets at Glasgow, packed, and shipped to Karachi for the final rail journey to Quetta, where the crates were assembled and sold to the dealers. The difficulty was not to sell the crates, but to make them up fast enough to keep an adequate reserve stock during the fruit season.

At the beginning of this work an interesting thing happened. After the crates had been designed and successfully used for my own consignments, the local traders without exception refused to adopt them. They only saw one side of this question: they did not see how much better and further my grapes travelled than theirs and how this increased the demand by bringing in distant places, which had only heard of the grapes of Afghanistan and Baluchistan. But the fruit dealers all over India soon insisted on their consignments being packed exactly as mine were. The demand for the improved crates then went up by leaps and bounds. It is safe to say that had this work been confined to the design of packages only and had it not included actual trading, by which the whole subject could be explored, no reform of the frontier fruit trade would ever have taken place.

But the most difficult obstacle of all was to persuade the Indian railways to unify their rules and to agree to return the empty fruit crates free of charge in return for the increased revenue which resulted from standardization. My proposals every year were duly placed before the

Fig. 3. Returnable and non-returnable crates for tomatoes

Railway Conference Association and were invariably rejected. Then suddenly, to my great astonishment, they were accepted in full.

This experience shows how necessary it is for the innovator in agricultural matters to have complete freedom for working out his ideas and ample time to get them adopted. It shows, also, how important it is for the scientist to keep his attention directed to every practical aspect of the problem before him, to neglect no detail, however humble. Nevertheless, these fruit-packing results would not have been possible, had not the grapes themselves been well grown. The length of the life of the grape after harvest is a short one unless a suitable variety is grown and the details of the actual growing are correct. This principle applies to most fruit and to most produce. Keeping power, like disease resistance, depends on the kind grown and on correct methods of agriculture.

But the most useful lesson in grape growing I learnt in Baluchistan must be mentioned last of all. I realized what a healthy vine should look like at all stages of its growth and how eloquent are the leaves, the buds, and the old wood about the soil conditions needed for ideal root development. How essential this item of my education has been will be evident from what follows.

My next lesson in the cultivation of the vine was in Africa—in Cape Colony in the spring of 1933 and in Algeria and Morocco in 1936. Generally speaking, all the vineyards I saw were only moderately affected by disease. But nowhere were vines to be seen with quite the same health and vigour as those on the Western Frontier in India. I put this down at the time to a want of balance between the vines and the livestock. Everywhere were large areas under vineyards, but there did not seem to be anything like enough farmyard manure. But a change is now taking place in the Western Province of South Africa. Even in 1939 the vine growers were beginning to take up the Indore Process. One such example on the main road between Somerset West and Stellenbosch was referred to by Nicholson in the South African *Farmer's Weekly* of 23rd August 1939 in the following words:

"Motorists travelling along this road cannot help noticing how healthy this farmer's vineyards look and how orderly is the whole farm. Early this winter I visited it in time to see the huge stacks of manure—beautiful, finely rotted bush, which had been helped to reach that state by being placed in the kraal under the animals. Pigs had played their part too. During the wine-pressing season all the skins of the grapes are

136

fed to the pigs and later returned to the vineyards in the form of manure."

Since these words were written South Africa has become compost-minded and I am informed that much more attention is now being paid to livestock as a factor in successful grape growing and to the systematic conversion of all available vegetable and animal wastes in humus.

In Algeria and Morocco every available acre seemed to have been planted in vines, but the supplies of farmyard manure seemed to me to be quite inadequate. The methods of grape growing, the prevention of disease, and the manufacture of wine closely followed those in the south of France, which I was soon to study in some detail.

My last course of instruction in the raising of grapes took place during the summers of 1937, 1938, and 1939 in the Midi, where in the course of many memorable tours in the company of the late Mr. George Clarke, C.I.E., a former colleague in India, I saw many thousands of acres under the vine and learnt a good deal about the way this crop is cultivated in the south of France. What struck me most, besides the shortage of farm-yard manure, was the vast sums of money spent on artificial manures to grow the crop and on poison sprays to keep the various fungous diseases at bay. In spite of all this, the crop did not seem at home. The foliage in particular looked wrong. Almost everywhere in the areas given up to vineyards there seemed to be far too little farmyard manure. In one large group of vineyards near the mouth of the Rhône, where tractors had almost entirely replaced the horse and artificials were relied on for growth, I never saw the spraying machine and the poison spray so much in evidence. One interesting result of all this was that the grapes pro-duced in these vineyards could no longer be used to make wine, but were devoted to the production of alcohol for diluting the petrol needed for motor-cars. No one, however, seemed to realize the significance of all this—the complete failure of artificials to maintain health in the vines and quality in the produce.

A sharp look-out was kept during these tours for vineyards in which the appearance of the foliage and of the old wood should tally in all respects with those of Central Asia, namely, well-grown plants looking thoroughly at home and in which the wood, the foliage, and the young grapes possessed the bloom of health. At last, near the village of Jouques in the Department of Bouches du Rhône, such vines were found. They caught my eye on the left-hand side of the road, as our car slowly de-scended by a winding roadway from the high ground above to the valley

137

below. We halted and made discreet inquiries. These vines had never received any artificials, only animal manure; the vineyard had a local reputation for the quality of its wine. Arrangements were then made with the proprietress to have the active roots examined. As was expected, they exhibited the mycorrhizal association. The vine proved to be a mycorrhiza former. The perfect nutrition, the high quality, and good keeping power of the grapes, the long life of the variety, and the absence of disease in Central Asia were at once explained. It was equally obvious that the general degeneration of the vineyards of the Midi and the need for poison sprays to keep fungous diseases in check, as well as the necessity for the plant breeder to produce an endless supply of new varieties, could all be traced to failure to realize the vital importance of livestock and of real humus for this ancient crop.

Obviously, at some period in her history, France took the wrong turning in the cultivation of the vine and failed to realize the need of balance between livestock and crops. It is more than likely this change began with the increased demand for wine which followed the Industrial Revolution and the growth of the urban areas. In all probability the *Phylloxera* epidemic, which overwhelmed the vineyards towards the end of the nineteenth century, was the first of Nature's warnings and the beginning of the writing on the wall. More will come.

Looking at the cultivation of the vine from all possible angles and bearing in mind the lessons of the Orient, there can be little doubt that the faithful adoption of the law of return will speedily put an end to most of the diseases of this crop and, at the same time, establish a new base line for the investigations of the future. In the training of the investigators of to-morrow it seems essential that our future instructors should widen their experience and take into consideration the lessons the Orient has to teach us about the stability of the variety and its resistance to disease once the manuring follows the lead of Nature.

FRUIT

My active interest in the problems of fruit growing and the reaction of the fruit tree to disease began in the West Indies in 1899 and has continued ever since. From 1903 to 1905 a good deal of attention was paid to these matters while on the staff of the South Eastern Agricultural College at Wye. At Pusa I had a large fruit plantation under my charge for nineteen years and spent a good deal of time in the study of the problems underlying fruit production. This included an investigation of the factors concerned in the effect of grass on fruit trees. The work involved the detailed examination of the root systems of a number of different species throughout the year and the way the trees and the soil came into gear. The results of ten years' work were summarized in Chapter IX of *An Agricultural Testament*. At Quetta on the Western Frontier I was provided with a small experiment station from 1910 to 1918, where fruit was the main interest. On retirement in 1931 I continued my studies of fruit problems in my small garden at Blackheath. My experience of fruit and its diseases has, therefore, extended over a period of forty-five years.

During this period a few very interesting cases both of loss of quality and of active disease have been investigated, the results of which are now set forth in chronological order.

The first of these problems was met with at Pusa in the case of the peach. Quite by chance one of the peach plots happened to be planted on a well-drained, permeable soil, in which the growth was far above the average of the locality. The yield and quality of the peaches were outstanding. It was quite easy to remove the skin of any of these ripe peaches in one piece—a quality test as good as any. On several occasions towards the end of the crop the weather changed—the dry, hot, westerly winds, usual during the ripening period, gave place to the damp, easterly winds which always precede the south-west monsoon. With this change in the humidity two things always happened: (1) the peaches lost their quality and became tasteless; (2) they were then attacked by the fruit-fly. Now these fruit-fly attacks never occurred while the air was dry and the fruit retained its taste and quality. No sooner had the damp winds destroyed the flavour than the fruit-fly appeared and its maggots proceeded to devour the crop. Even if it had been possible to keep the fruit-flies in check, nothing would have been gained for the simple reason that when the quality is lost peaches are hardly worth saving.

Another interesting thing happened at Pusa in connection with the peach. The raising of quality crops depended on an ample supply of irrigation water after the fruit had set, because during this period little or no rain was received, the upper soil was dry, and the extensive surface root system of this crop remained dormant unless kept moist by irrigation. With no irrigation the peach managed to survive the hot season and to ripen a small crop, but with this difference—the peaches were

139

small, hard, and quite devoid of quality. The explanation appears to be this. The peach tree, like the other fruit trees under study at Pusa, has two root systems—a well-developed, surface system, which comes into action during the growth period provided the surface soil is moist enough; if the peach is irrigated during the hot season, these surface roots begin to function when the buds open in the spring and continue

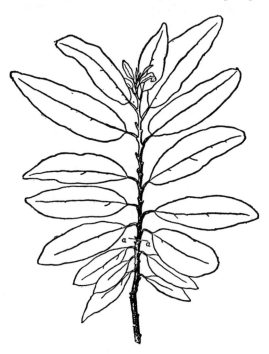

Fig. 4. Hot weather (below *a a*) and monsoon foliage (above *a a*) of the custard apple

in action during the rains, till the leaves fall; if, however, the trees are not watered, the surface roots remain dormant till the south-west monsoon in June. The function of the deep root system is to maintain the water supply during the hot season, and for this purpose new absorbing roots are produced every hot weather in the deep, moist layers of soil down to twenty feet from the surface. Obviously the two different methods of supplying the peach with water lead to very different results as regards the quality of fruit. These two methods also affect the leaves as well. Under irrigation, large, well formed leaves of the right colour

140

were produced throughout the season: there was no difference between hot weather and rains leaves. But when the trees relied for water on the deep roots only, the hot weather leaves were small and pale green, changing suddenly into large, dark green leaves when the monsoon in June brought the surface roots into action. Unfortunately I did not have these leaf differences recorded in drawings in the case of the peach, but only in the custard apple, where the results were closely similar (Fig. 4).

These facts suggest a promising direction for the study of quality in fruit. The development of quality depends entirely on surface roots and on the food materials these roots collect. As the peach is a mycorrhiza former and as this relationship occurs only in the surface roots, we have in this species and the other fruit trees cultivated in India, all of which possess two root systems and all of which are mycorrhiza formers, perfect instruments for breaking new ground in nutrition and in the detailed study of the fungus-root partnership.

Another very good example of a tropical fruit, in which the mycorrhizal association affects the upper of two root systems, superficial and deep, and thus plays an important part in the development of quality and in disease resistance is the guava; this fruit is easily grown and cultivated. This root development is shown in Plate IV. Further details of the investigations made at Pusa on this crop will be found in *An Agricultural Testament*, Chapter IX.

Another of the crops I grew at Pusa was the banana. When manured with farmyard manure, the response to this treatment as regards yield and quality was amazing. So it is when leaf-mould from the forest is used, as I once observed in the Botanical Station at St. Vincent in the West Indies about 1900, when some suckers of various varieties imported from India were tried out. The effect of leaf-mould was to confer on the fruit flavour and quality otherwise unknown. Further, both at Pusa and St. Vincent there was not the slightest trace of disease.

How very different are the plantation results in the West Indies and Central America, where large areas of steep hillsides under forest have been converted into banana fields. As long as the original humus made by the trees lasts, all goes well, but the moment this is exhausted one fungous disease after another makes its appearance and does great mischief. It appears that in these modern plantations little or no provision has been made for livestock and the preparation of large quantities of compost for maintaining the soil in a fertile condition.

That this is needed is suggested by the fact that the banana is a mycorrhiza former.

That properly made humus will always be essential in banana cultivation is suggested by the following extract from a letter dated 27th February 1944 from a correspondent in Southern Rhodesia, Mr. A. D. Wilson, Burnside, Bindura, who has been trying out the effect of humus on various fruit trees. As regards the effect of humus on the banana, he writes:

"*Bananas.* The effect of compost on these has been perhaps the most marked of anything I have done. Bananas are not considered a commercial proposition in Southern Rhodesia and for four years I worked away without using compost. Then I began to apply it—the change was remarkable. Year by year the plants grew larger, the bunches increased their yield till to-day I can expect bunches that carry 200 large bananas and more, and have a flavour better, so the Chief Horticulturist says, than any imported article."

Another interesting example of the effect of organic manures on the orange has just come from the Mazoe valley in Southern Rhodesia. In a letter dated 9th June 1944 Captain Moubray writes:

"I have been watching an orange grove belonging to one of the large companies. It is about twenty years old and has been fed all its life on little but artificials containing a large percentage of sulphate of ammonia. It is just about finished—the trees are full of dead wood and the crops it bears are now unprofitable—the soil is practically dead. Opposed to it is another grove further down the Mazoe valley, which has to a large degree been fed on organic wastes—it is still healthy and bears good crops. Again another one, which was chemically fed till a few years ago—the trees were cut off about four feet high and the treatment changed to organics—the trees are now coming away strong and healthy. I think I shall write a short article about it and call it 'Two Orange Groves.' [1] From all information I get the same thing has happened with tea."

As already stated, fruit was my principal preoccupation during the nine seasons, 1910–18, which I spent at Quetta in Baluchistan. Many further observations were made, some of considerable interest. The way in which green-fly attacks could be induced or checked at will on the peach and the almond is described in *An Agricultural Testament*, p. 164.

[1] This article appeared in the issue of *The Fertilizer, Feeding Stuffs and Farm Supplies Journal* of 15th September 1944.

Green-fly was unknown in the area under my charge until over-irrigation produced a heavy attack which was completely checked by restoring the aeration of the soil. This has been one of the neatest examples which has come under my observation of the effect of soil aeration on the health of a crop: the results were so well marked and so definite, two quite distinct foliages being produced, one fly-infected at the base, and one quite normal and free from infection further along the shoots. It was particularly noticeable that the fly did not spread from the infected leaves to the normal. The original purpose of the extra irrigation had been to try to store the precious irrigation water during the winter in the soil itself instead of allowing it to run to waste. Evidently Nature did not agree to this suggestion and showed her refusal in the usual way.

Among my most successful attempts to grow fruit at Quetta must be mentioned outdoor tomato growing; this had also been carried on at Pusa. Each plant was allowed to produce two stems which were tied to an ordinary wire fence of the right height, the tomatoes making a wall of foliage and fruit without any loss of space. The only manure used was cattle manure, but great trouble was taken to raise really strong seedlings for transplanting. Not only were the yield and quality far above the average, but the carrying power of the fruit was amazing. It was possible to send tomatoes from Quetta to the distant Calcutta market during August and September in ordinary railway vans, first through the terrific heat of the Sind desert, followed in the Gangetic plain by the moist, hot conditions of the Indian monsoon. The tomatoes arrived without damage or loss of quality, a fact I attributed to the care expended in their growth. Besides their keeping power and good quality, not the slightest sign of any insect, fungous, or virus disease appeared in these large-scale trials.

With this experience in retrospect, I was naturally intensely interested in a letter I received some years ago from Mr. A. R. Wills of the Tadburn Nursery, Romsey, in Hampshire. Mr. Wills asked my advice about the disposal of a considerable quantity of tomato haulm which had been attacked by the common wilt disease. I advised composting and returning the compost to the same houses for the next crop. This suggestion was somewhat violently opposed by one of the experts connected with the Ministry of Agriculture, who foretold dire results if my unorthodox proposals were accepted. Mr. Wills, however, decided to adopt them. The result was a fine crop, free from disease. Mr. Wills then proceeded to install the Indore Process at Tadburn and in this work was enthusiastically backed up by the foreman in charge. The result is that since those days Tadburn has never looked back and has gone from strength to strength.

Since 1934 in my small garden at Blackheath I have conducted an experiment to ascertain the effect of a fertile soil on the incidence of fruit diseases. When the garden was taken over in 1934 the acid, sandy soil was completely worn out and the fruit trees—apples, pears, cherries, and plums—were literally smothered by insect and fungous pests. They were the kind of trees that most people would have consigned to the bonfire. But instead they were carefully preserved and steps were taken to convert all the available wastes of the garden into humus. Some of this was given to the trees and the reaction of the pests to the new manurial treatment noted. Nothing very much happened the first year. The next year infection was noticeably less. The third year most of the pests had disappeared of their own accord, except in one case—a rather delicate apple tree, badly infested with American blight. During the fourth year this infection disappeared, but the tree is nothing like so robust as the others and again (1944) after a three years' abstinence from annual dressings of compost shows a distinct tendency to welcome a leaf disease —in this case due to a fungus. It may be that the stock on which this apple is grafted does not suit the sandy soil or that the combination of stock and scion is not a happy one. But, with this interesting exception, all the fruit trees have thrown off their pests and produced fruit of really exceptional size, quality, and keeping power. A small and rather old pear tree, which in 1934 was literally alive with green-fly and plant lice, armies of the latter being observed climbing up the stem, a really disgusting sight, has been restored to health: *the tiny, hard, uneatable pears of 1934 have developed into fruit of remarkable size and quality.* The twigs and leaves are now healthy and quite free from pests. No fungicides or insecticides were at any period used in this work.

Perhaps the most interesting experiment in this Blackheath garden concerns a common virus disease of strawberries. This arose out of a visit to the strawberry area round Botley, near Southampton, which, as is well known, has fallen upon evil days. The crop is grown by small-holders, but no provision was made for livestock and the production of animal manure. Substitutes, mostly composed of artificials, were used instead. *As long as the original stores of humus in the soil lasted,* all went well and a prosperous industry developed. Trouble then began. The soils lost their texture and permeability, and the strawberry plants began to be affected by virus and other diseases and then to go on strike. The area under crop dwindled. During the same visit I saw a large, well conducted strawberry farm near Southampton, on which farmyard manure was always applied. The crops were excellent and no soil troubles or pests were to be seen. I secured samples of the roots of these thriving strawberry plants, and asked Dr. Rayner to examine them. As I expected, the strawberry is a mycorrhiza former and therefore likely to respond to properly made humus.

At this point I began to wonder what would happen to virus-infected strawberries, if they were grown in compost. Would the affected plants recover? If virus-free and virus-infected plants were grown in compost side by side, would any infection take place? What would be the result of starting a new plantation in heavily composted soil from runners, half of which came from the virus-infected plants and half from healthy plants? Accordingly such a plantation was made. Two samples of Royal Sovereign strawberries were secured—one from an experiment station, certified to be attacked by virus, the other from the best commercial strawberry farm I knew of in England, where no virus had occurred. The plots were arranged side by side on land well manured with compost. The results were interesting. No infection of the healthy strawberries occurred: the virus-afflicted plants recovered: the new plot from equal numbers of runners from the original plantings was free from any trace of disease and, moreover, has yielded good crops of fine quality. The virus disease of strawberries appears, therefore, to be a mare's nest and to result from methods of farming which are inadmissible. The remedy is to combine livestock with strawberry growing and to convert all the vegetable and animal wastes into humus.

It occurred to me in the course of this work that the Southampton strawberry industry could be assisted or perhaps salvaged outright if use could be made of the large quantities of unused humus in the controlled tips near the city. I visited one of these controlled tips near Bitterne and found, as I expected, that it was a veritable humus mine. All that was needed was to separate, by simple screening, the refractory material and to place the resulting humus at the disposal of the strawberry growers. But all my efforts to get this done failed to overcome the inertia of departmentalism. The municipal authorities concerned with the tips and the county authorities anxious to help the strawberry industry were widely separated and independent bodies. I could not, in the brief time at my disposal, discover the secret by which the various bodies concerned could be brought into fruitful co-operation. In the meantime, the strawberry industry continues to decline. This episode reminded me of the anecdote recounted in Thackeray's *Book of Snobs,* where the King of Spain was burnt to death because no Director of Etiquette was available to set the machinery of the Court into harmonious and effective action, so that one of the footmen on duty could pour a nearby bucket of water on the unfortunate monarch.

While in Westmorland (1940–3) I saw an excellent example of recovery from virus disease, by means of compost, in raspberries at Levens Hall. Twelve years ago Mr. F. C. King, the head gardener, decided to put to a crucial test the current views on the running out of varieties and to discover whether this is due to improper methods of soil management or to a real breakdown in constitution. For this purpose he started in fertile soil a new raspberry plot from the most virus-infected stock of Lloyd George he could find. The plants soon made a complete recovery from virus. I saw them in 1943 and found them free from disease and still producing heavy crops of fine fruit, quite up to exhibition standard. This was one of the best examples of *the retreat of virus before soil fertility* I have so far seen.

In all these adventures in fruit growing I never had occasion to use a spraying machine for destroying a parasite, or any fungicides, insecticides, or germicides. *Disease resistance was left to the plant.* The only damage from parasites that could be regarded as at all serious were the attacks of peach fly at Pusa towards the end of the crop in those seasons when the moist currents which heralded the south-west monsoon caught the crop and destroyed its quality and made it attractive to the pest. Against accidents of this kind there can be no remedy—they must be accepted as inevitable. This long experience of the power conferred on the fruit tree by proper methods of manuring and soil management has helped to confirm my earlier ideas that bad farming and gardening are at the root of disease and that *the appearance of a pest should be regarded as a warning from Mother Earth to put our house in order.*

There is a further point to consider. If fruit trees need to be drenched with poison sprays before they can produce a crop, what is the effect of such fruit on the health and well-being of the people who have to consume it? We know these practices kill the bees and also the earthworms.

TOBACCO

One of the crops under study at Pusa between the years 1905 and 1923 was tobacco grown for leaf and also for seed. Only one disease, which resulted in malformed dwarf plants, was met with during these nineteen years. This trouble has since been proved to be due to virus. Such affected plants were quite common in the various cultures for the first two years, then they became fewer and by 1910 had disappeared altogether. Similar diseased plants occurred in the neighbourhood in the fields of the cultivators from whom a portion of the labour force was obtained. At no period were any steps taken to control this disease or to regulate the movements of the labourers. Nevertheless, no infection was spread or was carried *once correct methods of growing tobacco were adopted.* These consisted in raising the seed on humus-filled soil, careful attention to the surface drainage, and organic manuring of the nurseries, the production of well-grown material for transplanting, and the growth of the leaf tobacco on soil fertilized by various organic manures including farmyard manure. At no period in these nineteen years was the soil of the tobacco nurseries sterilized nor were artificials or spraying machines used. My tobacco cultures, which always earned the respect of all who saw them, were examples of organic farming pure and simple. Once the details of tobacco growing were mastered there was no disease of any kind: *the plants protected themselves* against every form of parasite as well as virus.

Captain Moubray informs me that similar results are now being obtained in Southern Rhodesia, where tobacco is an important commercial crop. The replacement of artificials by freshly prepared compost in the nurseries and in the tobacco fields was at once followed by a very marked diminution of virus trouble.

That the other tobacco diseases which of late years have begun to trouble the farmers in Rhodesia are due to an impoverished soil is suggested by the appearance of eelworm in this crop. This disease and its prevention are referred to in the *Rhodesia Herald* of 4th September 1942 as follows:

"At Darwendale, Mr. O. C. Rawson has applied five tons of compost per acre to infested tobacco land. In the first year there was a reduction of eelworm, and in the second year, without a further application, the eelworm disappeared. Other tobacco farmers began to report similar experiences. The compost, of course, was applied for its fertilizing value and the consequences on the eelworm population were a surprise."

Tobacco has not proved to be an exception to the long list of crops which are mycorrhiza formers. Samples of the surface roots of Rhodesia tobacco, taken from plants grown by means of freshly prepared humus, exhibit, as was expected, *this very significant symbiosis.* It is more than probable that quality in this crop will be found to depend, among other factors, on the efficiency of the mycorrhizal association. If this proves to be the case, the restoration of high quality in the cured product in places like Cuba will not be a very difficult matter once properly made humus replaces artificial manures.

LEGUMINOUS CROPS

The leguminous crop as a rule is very sensitive to soil conditions and in particular to poor soil aeration and its consequences. In the course of the current work at Pusa and Indore some interesting cases of the relation between soil conditions and disease occurred in these crops.

Perhaps the most interesting was one which was repeated year after year at Pusa in the case of a vetch—*Lathyrus sativus,* L.—known as *khesari.* The various unit species of this crop, collected from all parts of India, were grown in pure culture in small oblong plots about fifteen feet by six feet. Infection by green-fly occurred every year on a number of these cultures, but the trouble never spread to the remainder. The plots could be divided as regards infection into three classes: plots immune to green-fly; plots lightly affected; plots heavily attacked. Careful note of this infection was made and the cultures were repeated year after year. The same results were invariably obtained. On looking up the history of these cultures, it was found that the immune types came from the Indo-Gangetic alluvium, the heavily infected unit species from the black cotton soils of Peninsular India, the moderately infected types from the region near the Jumna, where the transition soils between the black cotton soil area and the alluvial tracts occur. The root system of these three sets of types was then explored. It was found that the immune cultures had superficial roots; those heavily infected had very deep roots; the slightly infected types had root systems intermediate between the two. These observations suggest that defective soil aeration, particularly affecting the deep-rooted varieties, was at the root of this

148

CUTTING AGERATUM

COMMUNAL COWSHEDS

PLATE IX. COMPOSTING AT GANDRAPARA

CRUSHING WOODY MATERIAL BY ROAD TRAFFIC

SHEET-COMPOSTING OF TEA PRUNINGS

PLATE X. COMPOSTING AT GANDRAPARA

green-fly infection, a view which has frequently been confirmed since these observations on *khesari* were made.

Another interesting case of disease in a leguminous crop occurred at Indore in a small field of gram (*Cicer arietinum*) about two-thirds of which was flooded one day in July due to the temporary stoppage of one of the drainage canals which took storm water from an adjacent area through the estate. A map of the flooded area was made at the time. In October, about a month after sowing, the plot was heavily attacked by the gram caterpillar, the insect-infected area corresponding exactly with the inundation area. The rest of the plot escaped infection and grew normally. The insect did not spread to the other fifty acres of gram, grown that year alongside. Some change in the food of the caterpillar had obviously been brought about by the alteration in the soil conditions caused by the temporary flooding.

Perhaps the most interesting case of the relation between soil conditions and disease which I observed occurred at Indore in the case of a field of *san* hemp (*Crotalaria juncea*, L.) intended for green-manuring; this, however, was not ploughed in, but was kept for seed as the growth seemed so promising. But after flowering the crop was smothered by a mildew; no seed was harvested. To produce a crop of seed of *san* on the black soils I had to copy the methods of the cultivators who always manure this crop with farmyard manure when seed is required. Instead of farmyard manure, I used compost the next year. No infection with mildew took place and an excellent crop of seed was obtained.

It is more than probable that this observation applies to leguminous crops generally. Whenever they are grown for seed, the best results are likely to be obtained with compost or farmyard manure. In olden days it used to be the custom to muck leguminous crops like clover, but the practice was given up after the role of the root nodule in fixing atmospheric nitrogen was discovered. But the root nodule is only a device to save these crops from nitrogen starvation. Nodules by themselves are not sufficient for the rapid growth and maturation involved in producing a full crop of seed.

Confirmation of the view that humus is needed by the leguminous plant if heavy crops of seed are to be obtained is coming to hand. In this country, in the case of clover, Mr. R. G. Hawkins, Lightwaters, Panfield, Braintree, Essex, in a letter dated 30th May 1942, reported:

"For some years now I have inspected crops of Essex red clover and I have noted that the yield of seed is invariably higher on those farms which keep stock, so that the land receives a periodic dressing of dung. The difference is most pronounced in those years when clover seed is generally a poor crop." (*News-Letter on Compost*, No. 6, 1943, p. 56.)

In Southern Rhodesia Captain Moubray, in a letter dated 1st June 1942, commented on the outstanding yields of *san* hemp seed he had obtained on composted land *in a bad season. He obtained no less than three times the average yield of his neighbourhood* (*News-Letter on Compost*, No. 4, 1942, p. 37). In a recent letter to the South African *Farmer's Weekly* of 7th June 1944, he writes:

"I remember, years ago, Sir Albert Howard telling me that the virtue of properly made compost lay not only in its contribution of humus, but also in its work as an inoculant. He suggested that its application in comparatively small quantities, before planting a legume, would considerably increase the seed yield of such a crop. I have found this to be so."

Here is a subject which urgently needs detailed study. We know that the large group of leguminous plants are mycorrhiza formers. It may well be that the efficiency of this association is one of the chief factors in seed formation. But whatever the explanation may be, it is clear that our fathers and grandfathers were right when they mucked the leguminous crop and that the agricultural colleges are wrong in telling the farmers that the root nodules will look after the nitrogenous manuring of these crops.

POTATO

My study of the potato crop only began at Quetta during the war of 1914–18 in connection with the drying of vegetables for the troops on active service. At first a supply of potatoes was purchased from the neighbouring tribesmen, but these proved unsuitable as the slices turned black in the drying process. This appeared to me to be due to the excessive quantities of irrigation water used and to the subsequent caking of the soil round the tubers. An area of potatoes was then grown at the Quetta Experiment Station, taking care to use the minimum amount of water applied to the roots only, leaving the earth of the ridges where the tubers were formed quite dry. The result was that no more blackening of the slices occurred. Soil aeration is obviously a factor in successful potato growing.

My second contact with this crop occurred in the Holland Division of Lincolnshire, where for some three years (1935–8) I was provided with ample facilities for study by the late Mr. George Caudwell on his farms near Spalding in connection with an investigation on green-manuring. On these farms the supply of farmyard manure was quite insufficient for the large area—some 1,500 acres—under potatoes. Heavy dressings of a complete artificial were then the rule.

Two common potato diseases were observed and studied in South Lincolnshire—blight and eelworm. In damp, close weather potato blight always occurred and had to be kept at bay by repeated dustings with finely divided copper salts. This disease was much more prevalent on the popular King Edward variety than on Majestic. I was asked why this was so. It appeared to me that the answer would be found if the root systems of these two varieties were compared. King Edward has a much deeper root system than Majestic and would, therefore, the more readily suffer from poor soil aeration, particularly during a spell of damp, close weather which would make the surface soil run together into a crust. Not only was the root system of Majestic markedly superficial, but the roots showed well defined aerotropism and invariably left the soil and grew on the surface under the fallen potato leaves.

I then went into the history of the celebrated potato area south of the Wash and found that some sixty years ago it was under grass. When first ploughed up for potatoes, the land was so rich in humus that crops sometimes as high as twenty-five tons to the acre were obtained. At first potato blight was unknown. But as the humus in the soil became worn out, dressings of superphosphate were first needed to keep up the yield, then the potato blight made its appearance, followed by the spraying machine, the poison spray, and the use of artificial manures, the annual applications of which gradually increased till they have reached fifteen hundred weight to the acre or even more.

These facts suggest that the real cause of potato disease is not, as is supposed, the potato blight assisted by hot and damp still air, but worn-out soil. This view could easily be tested by bringing up, by means of compost, one or two farms in South Lincolnshire to a fertile condition, comparable with what they were some sixty years ago when the pastures were first brought under potatoes. Would the potato on such fields be attacked by blight even if it had no assistance from poison sprays? Judging from what happens in our best walled gardens, in which good old-fashioned muck is the rule and in which artificials are never used, I think the answer would be in the negative.

That potato blight is of no consequence if ample farmyard manure is used to raise the crop and the plants are not grown too close together is proved by the experience of Mr. John Tarves at Heversham in South Westmorland. In 1943 I visited this garden and was shown a large potato plot on well-drained land facing south and protected from wind on the east and west, which was kept in good condition by farmyard manure and on which potatoes had been grown continuously for forty-five years. The rainfall at Heversham is very high and well distributed, the amount of sunshine is much below that of South Lincolnshire, and at first sight one would expect that here ideal conditions for potato blight had been provided. Nevertheless, in this garden this disease had caused no trouble and preventive spraying was unknown.

I spent some time in the Spalding area in the study of the eelworm disease of potatoes. This is caused by the invasion of the roots by a species of eelworm which dwarfs the plant and prevents the formation of even a small crop. Eelworm is a comparatively recent disease in this area and, as a rule, first appears on the high, light land. *In such cases a remarkable change in the flora and in the soil structure precedes the outbreak.* The weeds are those of semi-waterlogged and badly aerated soils and include the mare's tail, a species of *Equisetum,* known locally as toad-pike. The soils have lost their texture, the compound particles their cement, and the blue and red markings characteristic of heavy, clay subsoils have made their appearance. This condition is the result of continuous dressings of stimulating fertilizers which lead to the destruction of *the humic cement* needed to maintain the compound soil particles. The appearance of eelworm in these potato soils is the writing on the wall and marks the complete failure of the present fertilizing practice, the replacement of farmyard manure by artificial manures. No further potato crops are possible till the tilth and fertility of the soils have been re-created.

As is usual in such cases, the experts were busy at the wrong end of the problem. The life history and activities of the eelworm were being studied and all kinds of methods except the right one were being tried to destroy the parasite and to stimulate the crop to ward off the disease. The result has been a complete failure. No one has seemed to grasp the fact that eelworm is one of the frequent consequences of poor soil aeration and that the cause of the trouble must be sought in a critical examination of farming practice. *This pest is one of the results of upsetting the balance between arable and livestock* and *trying to find, by means of chemistry, a substitute for good old-fashioned muck.* In all these eelworm outbreaks the soil's capital has been transferred to the profit and loss account. What will the reverse process cost before these lands are fully restored? Is the process a reversible one? If so, who is to meet the cost?

Confirmation of the views set out above that eelworm in potatoes is due to an impoverished soil comes from Southern Rhodesia. The first results are summed up in the *Rhodesia Herald* of 4th September 1942 as follows:

"Some years ago Mr. S. D. Timson, Assistant Agriculturist, noticed a garden where the vegetables were strong and healthy and the flowers bright and vigorous. He was surprised to learn that three years earlier cultivation had been almost abandoned because of the heavy infestation of eelworm. The excellent conditions he saw followed a good dressing of compost.

"He immediately began to observe the results of compost in regard to eelworm, to make practical tests, and induce farmers to experiment. Once the inquiry was begun evidence began to pour in."

That compost will prevent eelworm attacks on potatoes and other vegetables has again been demonstrated on a large scale at Salisbury in Southern Rhodesia by Mr. E. C. Holmes who, in the issue of the South African *Farmer's Weekly* of 14th June 1944, writes:

"Since I started using compost I have eradicated eelworm from my gardens, and I have no less than sixteen vegetable gardens spread all over my farm of 2,333 acres."

This eelworm story is being continued in Rhodesia. In the *Rhodesia Herald* of 7th July 1944, the following article appeared:

SATISFACTORY EXPANSION IN THE MAKING OF COMPOST

Tobacco Growers Report Excellent Progress from Its Use

"The expansion in the making and use of compost continued during 1943, states Mr. S. D. Timson, Government Agriculturist, in the course of his annual report. Tobacco growers, he states, gave compost much increased attention, and they continue to report excellent results from its use, and in particular that it gives better quality and greater freedom from disease. It also allows the rate of application of fertilizers to be much reduced without reduction of yield. Its use, as was to be expected, was not usually beneficial on virgin soil.

153

"Further reports were received from farmers that applications of compost to soil infested with eelworm resulted not only in good yields of tobacco and vegetables, despite the infestation, but also in the disappearance of the pest from the soil the year after the compost was applied.

"A striking example was in the vegetable garden of the Witchweed Demonstration Farm, where an extremely severe infestation was completely cleared up following an application of compost. On the same farm further evidence was recorded supporting Mr. Timson's previous reports of the beneficial effects of compost in controlling witchweed.

Well Satisfied

"In 1940 there were 674 farmers making compost; in 1943 the number had increased to 1,217. In the same years the amounts of compost made were 148,959 and 328,591 cubic yards (2 cubic yards = 1 ton).

"The largest producers had made from 4,000 to 9,700 cubic yards a year. The largest producers of fat cattle were now making compost instead of collecting kraal manure. They reported they were well satisfied with the change particularly in respect of *the elimination of weed seeds and* THE REDUCTION OF THE FLY NUISANCE."

There is another potato trouble in South Lincolnshire which is not caused by insects or fungi, namely, the loss of the power of reproduction. After two or three years the potatoes of one crop cannot be used to raise the next. The yield then becomes unremunerative and fresh seed has to be imported at great expense from outside areas like Scotland, Northern Ireland, or North Wales. As this loss of reproductive power develops, the cause is considered to be due to virus. Again the research workers are starting at the wrong end and are trying to find varieties immune to virus. The results so far obtained, as far as practice is concerned, are not impressive. Indeed, it would seem that this trouble is getting worse, as the efficiency of Scotch seed is said to be falling off. If this should continue, the Lincolnshire potato industry will find itself in difficulties. The fresh start every two or three years will no longer be possible unless some alternative supply of new seed can be found.

That these frequent changes of seed of any particular variety and indeed of the production of new varieties of the potato by plant breeding methods are both unnecessary, provided proper attention is paid to the maintenance of soil fertility by organic manuring, is proved by the experience of the islanders of Tristan da Cunha, that lonely settlement in the South Atlantic rarely visited by ships. Here changes of seed are out of the question on account of the inaccessibility of the island. In a letter, dated 15th March 1945, Major Irving B. Gane, the Secretary of the Tristan da Cunha Fund writes:

"As you rightly surmise, the islanders use seaweed. A belt of thick kelp extends round the island some 400–500 yards from the shore, and rough seas wash large quantities on to the beaches. This is collected by the islanders and used for their potato patches.

"I am satisfied that the islanders have no means of changing the variety of the potatoes grown, and it would be safe to assume that the seed has been retained from year to year during the hundred years or so of the island's occupation.

"I, and my father before me, organized the despatch of stores to the island, and although we have sometimes included supplies of vegetable seeds, we have certainly never sent out any seed potatoes."

The situation in Great Britain, though alarming, is not really serious. All that is necessary in areas like South Lincolnshire is to revise the current method of potato growing by a drastic reduction in the area under potatoes, so that the head of livestock—cattle and pigs in particular—can be increased, and large areas put under temporary leys and cereals. In this way the raw materials for systematic compost making will be available on the spot. As these reforms proceed, the amount of artificial manures can be reduced. When the stage is reached when artificials and poison sprays are no longer necessary, the restoration of these wonderful soils will have been achieved. After this the experience of the past can be made use of to test current practices. If these soils begin to respond to artificials, attention should be paid to the humus supply. If potato blight appears, the aeration of the soil needs attention.

In the course of these potato studies a number of root samples were examined for the mycorrhizal association. All the results were negative. I understand from Dr. Rayner that the ordinary cultivated crop does not show this relationship, but that it has been observed on potatoes in the hilly regions of France near the Spanish border. Has the potato in the course of years lost something, or was its original introduction imperfect? Do the wild forms of this crop in its mountain home in South America show the mycorrhizal association, or does this crop manage to absorb, by means of its very extensive root system, the digestion products of the proteins during the early stages in the mineralization of the bodies of the soil organisms? In due course answers to these questions will no doubt be provided. They are likely to have an important bearing on disease resistance in this crop and also on the power of the plant to reproduce itself.

SOME PARASITIC FLOWERING PLANTS

A few cases of disease in which the active agents are flowering parasites must be recounted.

The first of these occurred on four meadows on the farm near Bishop's Castle in Shropshire where I was born and where I spent my early boyhood. The parasite was the well-known yellow rattle (*Rhinanthus Crista-Galli*), which invariably attacked the grasses and considerably reduced the hay crop. I noticed at the time that a pasture alongside, on which cattle and sheep grazed, never had any of this parasite, but my studies at this period did not embrace this common example of a semi-parasitic flowering plant and its haustoria, which fasten on the roots of the grass. Some fifty years later, however, I discovered that some of the live wires in the farming community have found how to eradicate this pest. They turn the affected meadows into pastures for a couple of years, when the urine and dung of the cattle strengthen the grasses to such an extent that yellow rattle disappears altogether. As the grasses are mycorrhiza formers, we have here a most interesting problem awaiting investigation. Does the humus formed in the soil of pastures in the spring and early summer by the sheet-composting of the vegetable and animal wastes confer on the grasses, by virtue of this association, the power to resist the parasite? If so, is the increased resistance to disease nothing more than the efficient synthesis of protein, due to the passage into the leaves of the grasses of the digestion products of the protein of the mycelium of the mycorrhizal fungus? If, as seems likely, the answers to these two questions are in the affirmative, a great stride forward will have been made in establishing a scientific explanation of the relation between soil fertility and health.

During my Indian service I again came in contact with one of these flowering parasites of the grass family. This time a species of *Striga* was observed on the roots of the sugar-cane. The cultivators in India invariably got rid of this pest by manuring the affected crops with farmyard manure, after which the parasite disappears. Is the mycorrhizal association, which is known to occur in sugar-cane, involved in this matter? It would seem so.

After my retirement in 1931, in the course of the humus campaign in Southern Rhodesia I heard of the witch-weed (*Striga lutea*), one of the pests of maize (another mycorrhiza former) and its control by humus. This interesting discovery was made by Timson, whose results were published in the *Rhodesia Agricultural Journal* of October 1938. Humus made from the soiled bedding of a cattle kraal, applied at the rate of ten tons to the acre to land severely infested with witch-weed, was followed by an excellent crop of maize practically free from the parasite. The control plot alongside was a red carpet of this pest. A second crop of maize was then grown on the same land. Again it was free from witch-weed. This parasite will therefore prove a valuable soil analyst for indicating whether the maize soils of Rhodesia are fertile or not. If witch-weed appears, the land needs humus: if it is absent, the soil contains sufficient organic matter. Witch-weed will then be regarded not as a pest to be destroyed, but as a most useful soil assessor and land valuer—as the friend, not the enemy, of the farmer.

9

DISEASE AND HEALTH IN LIVESTOCK

Aʙᴏᴜᴛ ᴛʜᴇ ʏᴇᴀʀ 1910, after five years' first-hand experience of crop production under Indian conditions, I became convinced that the birthright of every crop is health and that the correct method of dealing with disease at an experiment station is not to destroy the parasite, but to make use of it for tuning up agricultural practice.

FOOT-AND-MOUTH DISEASE

If this holds for plants, why should it not apply to animals? But at this period I had no animals, my work cattle had to be obtained from the somewhat inefficient pool of oxen maintained on the Pusa Estate alongside, with the feeding and management of which I had nothing to do. I therefore put forward a request to have my own work cattle, so that my small farm of seventy-five acres could be a self-contained unit. I was anxious to select my own animals, to design their accommodation, and to arrange for their feeding, hygiene, and management. Then it would be possible to see: (1) what the effect of properly grown food would be on the well fed working animal; and (2) how such livestock would react to infectious diseases. This request was refused several times on the ground that a research institute like Pusa should set an example of co-operative work rather than of individualistic effort. I retorted that agricultural advances had always been made by individuals rather than by groups and that the history of science proved conclusively that no progress had ever taken place without freedom. I did not get my oxen. But when I placed the matter before the Member of the Viceroy's Council in charge of agriculture (the late Sir Robert Carlyle, K.C.S.I.), I immediately secured his powerful support and was allowed to have charge of six pairs of oxen.

I had little to learn in this matter, as I belong to an old agricultural family and was brought up on a farm which had made for itself a local reputation for the management of cattle. My animals were most carefully selected for the work they had to do and for the local climate. Everything was done to provide them with suitable housing and with fresh green fodder, silage, and grain, all produced from fertile soil. They soon got into good fettle and began to be in demand at the neighbouring agricultural shows, not as competitors for prizes, but as examples of what an Indian ox should look like. The stage was then set for the project I had in view, namely, to watch the reaction of these well chosen and well fed oxen to diseases like rinderpest, septicaemia, and foot-and-mouth disease, which frequently devastated the countryside and sometimes attacked the large herds of cattle maintained on the Pusa Estate. I always felt that the real cause of such epidemics was either starvation, due to the intense pressure of the bovine population on the limited food supply, or, when food was adequate, to mistakes in feeding and management. The working ox must always have not only good fodder and forage, but ample time for chewing the cud, for rest, and for digestion. The grain ration is also important, as well as a little fresh green food—all produced by intensive methods of farming. Access to clean fresh water must also be provided. The coat of the working animal must also be kept clean and free from dung.

The next step was to discourage the official veterinary surgeons who often visited Pusa from inoculating these animals with various vaccines and sera to ward off the common diseases. I achieved this by firmly refusing to have anything to do with such measures, at the same time asking these specialists to inspect my animals and to suggest measures to improve their feeding, management, and housing, so that my experiment could have the best possible chance of success. This carried the day. The veterinarians retired from the unequal contest and took no steps to compel me to adopt their remedies.

My animals then had to be brought in contact with diseased stock. This was done by allowing them: (1) to use the common pastures at Pusa, on which diseased cattle sometimes grazed, and (2) to come in direct contact with foot-and-mouth disease. This latter was easy, as my small farmyard was only separated from one of the large cattle sheds of the Pusa Estate by a low hedge over which the animals could rub noses. I have often seen this occur between my oxen and foot-and-mouth cases. Nothing happened. The healthy, well-fed animals reacted to this disease exactly as suitable varieties of crops, when properly grown, did to insect and fungous pests—no infection took place. Neither did any infection occur as the result of my oxen using the common pastures. This experiment was repeated year after year between 1910 and 1923, when I left Pusa for Indore. A somewhat similar experience was repeated at Quetta between the years 1910 and 1918, but here I had only three pairs of oxen. As at Pusa, the animals were carefully selected and great pains were taken to provide them with suitable housing, with protection from the intense cold of winter, and with the best possible food. Again no precautions were taken against disease and no infection took place.

The most complete demonstration of the principle that soil fertility is the basis of health in working animals took place at the Institute of Plant Industry at Indore, where twenty pairs of oxen were maintained. Again, the greatest care was taken to select sound animals to start with, to provide them with a good water supply, a comfortable, well-ventilated shed, and plenty of nutritious food, all raised on humus-filled soil. One detail of cattle-shed management was the provision of a floor of beaten earth, which is much more restful for the cloven hoof than a cement or brick floor. This was changed every three months, the dry, powdered, urine-impregnated soil afterwards being used as an activator in humus production, for which it proved most suitable. In this way it was possible to bank the spare urine under cover without loss by rainwash or fermentation.

A special feature of the food supply of the oxen was the provision of ample silage for the months March to June, when little or no grazing was available on account of the dry, hot weather. The silage was made from the locally grown tall millet, cut up by means of a portable chaff cutter driven by a 5 h.p. portable oil engine. The cut silage was filled into pits about four feet deep with sloping sides and an earthen bottom for drainage. To prevent the infiltration of air into the mass from the surrounding earth the sides were leeped with a thick, moist, clay slurry just before filling. The cut silage was moistened by means of a sprinkler as it went into the pits, each of which was so designed that it could be filled with moist silage and covered in during one day's work. This is essential for the best results. It never pays to fill a silo bit by bit, as is so often the case in Great Britain. The centre of each filled silage pit was about eighteen inches above the ground level, the edges were flush with the undisturbed soil, a thin covering of dried grass was then applied, followed by a foot of earth. On the top of this earth covering were laid some heavy blocks of stone. All this consolidated the moist

160

silage and allowed the proper fermentation to begin. No additions such as molasses were ever used. Proceeding in this manner, excellent silage was obtained with practically no loss. Indeed, damage by percolating air was impossible, while the small amount of liquid produced was absorbed by the earth below. The size of each pit was so designed that it contained the silage ration of forty oxen for fourteen days. Seven of these pits were in use and they contained sufficient for an ample daily ration on the 100 days between March 8th and June 15th.

Besides the design of an efficient pit silo—than which nothing can be so cheap and effective—two other details are important. The transport of the silage from field to silo, the machines used in its preparation, as well as the strength of the average labourer must all correspond, otherwise a great waste of capital and of labour is bound to occur. Two Canadian oxen-drawn fruit trucks were sufficient to feed the small engine-driven chaff cutter, which just suited the labour. This modest outfit produced enough silage in the working day for 40 x 14, i.e. 560 rations.

Besides this silage ration during the hot months a little fresh green lucerne, raised under irrigation from heavily composted land, was given to the oxen almost every day.

The result of all this was a complete absence of foot-and-mouth and other diseases for a period of six years.

But this is not the whole of the foot-and-mouth story. When the 300 acres of land at Indore were taken over in the autumn of 1924, the area carried no fodder crops, so the feeding of forty oxen was at first very difficult. During the hot weather of 1925 these difficulties became acute. A great deal of heavy work was falling on the animals, whose food consisted of wheat straw, dried grass, and millet stalks, with a small ration of crushed cotton seed. Such a ration might do for maintenance, but it was quite inadequate for heavy work. The animals soon lost condition and for the first and last time in my twenty-five years' Indian experience I had to deal with a few very mild cases of foot-and-mouth disease in the case of some dozen animals. The patients were rested for a fortnight and given better food, when the trouble disappeared never to return. But this warning stimulated everybody concerned to improve the hot-weather cattle ration and to secure a supply of properly made silage for 1926, by which time the oxen had recovered condition. From 1927 to 1931 these animals were often exhibited at agricultural shows as type specimens of what the local breed should be. They were also in great demand for the religious processions which took place in Indore city from time to time, a compliment which gave intense pleasure to the labour staff of the Institute.

This experience, covering a period of twenty-six years at three widely separated centres—Pusa in Bihar and Orissa, Quetta on the Western Frontier, and Indore in Central India—convinced me that foot-and-mouth disease is a consequence of malnutrition pure and simple, and that the remedies which have been devised in countries like Great Britain to deal with the trouble, namely, the slaughter of the affected animals, are both superficial and also inadmissible. Such attempts to control an outbreak should cease. Cases of foot-and-mouth disease should be utilized to tune up practice and to see to it that the animals are fed on the fresh produce of fertile soil. The trouble will then pass and will not spread to the surrounding areas, provided the animals there are also in good fettle. Foot-and-mouth outbreaks are a sure sign of bad farming.

How can such preventive methods of dealing with diseases like foot-and-mouth be set in motion? Only by a drastic reorganization of present-day veterinary research. Instead of the elaborate and expensive laboratory investigations now in progress on this disease, which are not leading to any practical result, a simple preventive trial on the following lines should be started. An area of suitable land should first be got into first-class condition by means of subsoiling, the reform of the manure heap, and reformed leys containing deep-rooting plants like lucerne, sainfoin, burnet, and chicory, and the various herbs needed to keep livestock in condition. The animals should be carefully selected to suit the local conditions and should first of all be got into first-class fettle by proper feeding and management. Everything will then be ready for a simple experiment in disease prevention. A few foot-and-mouth cases should be let loose among the herds, the reaction of both healthy and diseased animals being carefully watched. The diseased animals will soon recover. There will most likely be no infection of the healthy stock. At the worst there will only be the mildest possible attack which will disappear in a fortnight or so.

Foot-and-mouth is considered to be a virus disease. It could perhaps be more correctly described as a simple consequence of malnutrition, due either to the fact that the proteins of the food have not been properly synthesized, or to some obvious error in management. One of the

162

most likely aggravations of the trouble is certain to be traced to the use of artificial manures instead of good old-fashioned muck or compost.

This long experience of foot-and-mouth disease suggests that an important factor in the prevention of animal disease is food from humus-filled soil. Three further questions suggest themselves. Does any supporting evidence exist for this view? Can the animal help us in our inquiries on disease prevention? Is disease due to causes other than those arising from an infertile soil? That the answer to all these questions is most emphatically yes will be clear from what follows.

SOIL FERTILITY AND DISEASE

One of the first pieces of supporting evidence was supplied in 1939 by the late Sir Bernard Greenwell at his estate at Marden Park in Surrey, where large quantities of Indore compost were made and applied to the land. Sir Bernard was a successful breeder of livestock, and after seeing the very striking results of compost on the crop naturally began to wonder what would be the effect of grain raised on composted land on his pedigree animals. For this purpose the effect of a grain ration, raised from soil manured with Indore compost, was compared with a similar one purchased on the open market on poultry, pigs, horses, and dairy cows. In all cases the results were similar. The animals not only throve better on the grain from fertile soil, but they needed less—a saving of about 15 per cent was obtained. The grain from fertile soil was found to contain a satisfying power not conferred by ordinary produce. But this was not all; resistance to disease markedly increased. In poultry, for example, infantile mortality fell from over 40 per cent to less than 4 per cent. In pigs, troubles like scour disappeared. Mares and cows showed none of the troubles which often occur at birth.

These Marden Park results are illuminating and should be carefully considered by investigators and particularly by statisticians. Hitherto in agricultural investigations special importance has always been paid to quantitative results—to yield in particular. But is this sound? If quality is as important as the Marden Park results indicate, yield is only of real significance when it includes quality. Quality, of course, does not end with the particular experiment. The produce affects the health and well-being of the animals and men who consume it. Such crops are, as it were, the beginning of a long chain of circumstances which must be followed to the end. If we stop at the yield, our work is obviously superficial. It may also be very misleading. Suppose, for example, two manurial treatments give the same result as regards yield, but the one produces A.1 quality, the other only C. 3. The statistician will say the experiment yields no significant result, because the weights are the same. The animal, however, will plump for the A.1 produce and the observant farmer will agree with the animal. The food of the animal is produce; the statistician feeds on numbers which can always be made to prove anything and everything.

Since 1939 a good deal of evidence in support of Sir Bernard Greenwell's results has been obtained. At Dry Clough Farm on the boulder clay near the town of Nelson in Lancashire, at an elevation of some 900 feet above the sea, the stock-carrying capacity of a hill farm has been raised from twenty cattle in 1910 to fifty-six in 1942, by means of sheet-composting with the help of liquid manure from the shippons spread systematically over the pasture. On this heavy clay land the formation of abundant humus under the turf has completely altered the botanical composition of the original herbage and has produced some first-class rye-grass pastures. The health of the cattle is now wonderful; milk fever has vanished; the animals are tuberculin tested and the herd is fully attested. The veterinary surgeon reports that it is the best T.T. herd he visits; there are no reactors. The financial results are equally satisfactory. Full details of this interesting case are to be found in the *News-Letter on Compost* (No. 4, October 1942, p. 4, and No. 6, June 1943, p. 32).

Lady Eve Balfour in *The Living Soil* (Faber and Faber, London, 1943) recounts some interesting results on her farm at Haughley in Suffolk with pigs. Pigs bred under modern housing conditions are very prone to the disease of white scour when they reach the age of about one month. If the attack is serious, it can cause considerable financial loss even if it does not actually kill the pigs. The text-books give the cause as lack of iron and recommend dosing with some iron preparation such as Parrish's Food, feeding such weeds as chickweed (which is rich in iron), or, as a third alternative, taking up pieces of turf and giving these to the young pigs. Lady Eve writes:

"I have made many experiments in connection with the curing and prevention of this trouble. From the turf remedy I tried experiments with ordinary soil from arable fields. It was not long before I found that soil gathered from a field rich in humus, where no chemicals had been applied, was quite as effective as turf, curing the pigs within forty-eight hours. Whereas soil from exhausted land, or land treated with chemicals, had no effect in curing the disease. I also noticed that young pigs running in the open on good pasture, provided it was not too hard for them to rootle (as, for instance, in hard frost, or very prolonged drought), never suffered from this disorder. It is never a menace to my herd now under any conditions, even in long spells of severe winter weather, when the ground is covered with snow, and the pigs have to be entirely housed up. Under such conditions I no longer wait for the first sign of scour, but regularly collect the soil of *fresh* mole hills, newly thrown up above the snow, on land I know to be fertile. Collected daily, this soil is friable in the hardest frost, and is equally good in very wet weather, for it is never sticky. The pigs eat it voraciously in incredible quantities, starting when about a week old. I sometimes add a little chalk to it, which the pigs seem to like."

As regards the housing of pigs, I often observed, while being shown over some of our modern piggeries, the obvious discomfort of the young pigs and their mothers condemned to lie on concrete floors with insufficient bedding. The sows always did their best to keep their family warm by lying crossways to cut off the draught. This might keep the pigs warm, but it would interfere with their air supply. Very young pigs have little or no hair for warmth; as they are close to the floor, it is imperative to give them enough fresh air or lung disease is certain. How far disease in young pigs is due to lying on cold concrete I cannot say, but I feel sure that, if the sows and their families could be consulted about concrete floors, the nature and amount of their bedding, and the general design of the piggeries, some of our agricultural experts would begin to learn a great deal about the real wants of this interesting animal.

Perhaps the most convincing piece of evidence in support of the view that the best way of reducing the diseases of livestock to a minimum is proper care and feeding has been provided by Mr. Friend Sykes on his 750-acre farm at Chantry near Chute in Wiltshire. Chantry is situated on the escarpment of the South Downs overlooking Salisbury Plain; the general elevation is some 800 feet above the sea; the thin, poor soil, plentifully supplied with flints, overlies the chalk. Notwithstanding the fact that this area had been completely farmed out and was practically derelict, Mr. Sykes decided it could be transformed into an ideal area for breeding racehorses with the right type of bone and a dairy herd that could protect itself against disease. This has been accomplished in a few years by means of efficient cultivation, including subsoiling, the use of temporary leys containing deep-rooted plants as advocated by the late Mr. R. H. Elliot in his *Clifton Park System of Farming*, the use of the open-air system of milk production, the sheet-composting of the temporary ley by means of the droppings of livestock, and the reform of the manure heap, so that much more muck and much better muck can be produced. The result of all this on the livestock and on the land has been remarkable: diseases like tuberculosis, mastitis, and contagious abortion have practically disappeared; the livestock are fed solely on the produce of the farm; the stock-carrying capacity of this land is still on the up grade; no artificial manures are used; the yield of crops like wheat, barley, oats, hay, and so forth has increased by leaps and bounds.

A detailed account of these Chantry results will be found in Appendix D (p. 289).

CONCENTRATES AND CONTAGIOUS ABORTION

On several occasions I have come across serious outbreaks of contagious abortion in some of our best dairy areas and on farms where much had been done for the livestock. On inquiring I always found that the diet of the milking animals included large quantities of feeding cakes obtained from various oil mills, the compound cakes being made up of the residues of imported oil seeds reinforced by other materials to produce a food which would stimulate milk production. The excessive use of these cakes seemed to me to be quite unsuitable for the ruminant stomach, which is designed for abundant roughage and not for such concentrates as compound cakes.

When asked my opinion as to the best method of treatment, I invariably replied that the organism associated with this disease is only a mild parasite and will only infect the vagina if the cow is malnourished, and that the cure will be found in getting the soil in good heart to begin with so that it can produce the cereals, pulses, and linseed needed to reinforce properly grown grass, silage, and hay. Even if, for financial reasons, this is not possible in the case of milking animals, it is obviously essential for the breeding animal, which produces the future generation of heifers.

166

SELECTIVE FEEDING BY INSTINCT

A growing volume of evidence is being obtained which indicates how very useful the animal can be in investigations on nutrition. In place of the present-day elaborate investigations, carried out in laboratories by teams of scientists, animal instinct, if rightly used, will provide us with much reliable information of the first importance. A few cases which have recently come to my notice may be cited.

In the course of the late Sir Bernard Greenwell's grass-drying experiments, carried out before the war, the question of analysing the product was discussed and I was asked to recommend a suitable man for the work. I pointed to his herd of pedigree Guernseys and said they would give real information as to the quality and nutritive value of any two sets of samples, if the animals were allowed a free choice. The findings of the animal could then, for purposes of academic rectitude only, be submitted to any competent analyst who would provide a set of conventional figures. This was done. The verdict of the Guernseys was duly confirmed.

In a set of trials of artificials on grassland in a park near Kirkby Lonsdale in Westmorland carried out some years ago there was no appreciable difference in the weight of produce, so the experiments were discontinued by the artificial manure interests which had sponsored them. But on the removal of the fencing the preference of the grazing animal for the dunged plots was most striking. These were eaten down to the roots, while the chemically treated areas were left alone.

I verified the above observations in the case of six pastures in front of my residence near Haversham. All are first-class rye-grass pastures with nothing to choose between them as regards soil, aspect, or drainage. Nevertheless, in 1941 the sheep and cattle which had access to all six fields at the same time consistently neglected one of them, the grass of which was allowed by the animals to grow at will. This particular field alone of the six had received a large dressing of artificials.

One of the best judges of quality in food is the domesticated cat, whose fastidious reaction to its rations is well known. In *The Living Soil* Lady Eve Balfour recounts an interesting experience:

"Last winter I noticed that the farm cats refused potatoes boiled for the pigs, when these had been purchased from a grower who uses artificials, but that later in the season, when I started to use the small potatoes from our own land, grown with humus, the cats ate them with avidity."

Another interesting example of selective feeding from Norfolk is recorded by the Rev. Willis Feast in the *News-Letter on Compost* (No. 3, June 1942, p. 13):

"One young farmer told me that he grew swedes, some with and some without artificials. He fed the 'withouts' first, and when they were finished had the greatest difficulty in persuading his beasts to start eating the 'withs.' "

Two somewhat similar cases from Scotland have just been reported by Mr. James Insch: (1) After sampling a pasture once, in which artificials had been applied, a farmer failed to get the cows to enter the field again the next day, although assisted by a boy and a dog. (2) A Scotch farmer grew two samples of wheat, one with muck and the other with artificials, and was pondering how best he could have these two lots of grain tested. He got the results much more quickly than he expected. Rats broke into his granary and devoured the produce of the mucked field and left the other severely alone.

Examples such as these quoted do not, of course, conform with the standards deemed essential by the laboratory worker and by the statistician. Nevertheless, they are of the greatest value as indicators of results which are being obtained all over the world when organic farming is practised on a large scale. Many of the pioneers have already accepted them and are busy creating examples without end of what a fertile soil can do for the health and well-being of the livestock nourished thereon. Everything will soon be ready for the advocates of artificials, or artificials and humus, to take up land alongside these examples of organic farming and show what they can accomplish. The decision as to which is the better of the two kinds of farming will be duly delivered by Mother Earth herself. It can never be given by the lawyers on either side, who are certain to indulge in infructuous disputations designed to postpone any verdict. In South Africa the pioneers have for some time been waiting for such a trial. But an unexpected difficulty has arisen. The protagonists of artificials have so far declined the contest. Is it because they fear the result and have no stomach for a battle of which there will be no to-morrow? If their position is a sound one, what better advertisement for artificials could be found than a clear-cut victory over *these tiresome disciples of organic farming?*

HERBS AND LIVESTOCK

Besides the way the food of animals is grown, there is another important factor which urgently calls for investigation. This is the botanical composition of our meadows and pastures, and the part played by herbs in maintaining the health of the animal.

During the summer immediately before the present war I came in contact in Provence with the famous meadows of La Crau, which produce the hay consumed in the racing stables of France and which is sometimes sent as far as Newmarket. These meadows are irrigated by silt-laden water containing a good deal of impure carbonate of lime, taken from the River Durance, and yield as many as three or four crops of hay a year. I examined a number of these meadows in detail and took samples of the young active roots of the grasses, clovers, and herbs. All proved to be mycorrhiza formers. The texture of the soil was excellent with plenty of humus under the turf. I was very much impressed at the time by the high proportion of herbs in this hay. It often reached 30 per cent of the whole and I began to wonder how far the value of this hay was due to the herbs. Have we omitted an important factor in our investigations on grassland and on temporary leys in this country? What is the effect of the herbs on the health of the grazing animal? What herbs are found naturally in our most celebrated pastures in central and western England?

While pondering over these matters, I happened to read the following letter from Major Owen Croft, which appeared in *The Times* of 8th November 1943:

WAR POLICY ON THE FARM
GRASSLAND UNDER THE PLOUGH

Gains and Losses

To the Editor of *The Times*
Sir,

The recent correspondence in *The Times* tempts me to bring this question of permanent pastures into proper perspective.

It is my belief that experienced farmers are horrified at this destructive policy of the ploughing up of the fine permanent pastures—which take forty to fifty years to establish. I speak with thirty-eight years' experience of farming my own land in a district which has (or had) permanent pastures of the highest quality. All will agree that there are districts in the British Isles where the land and climate are both unsuitable for the establishment of permanent pastures; and that in these districts, which include the higher sheep lands of Wales, farms have benefited enormously from the policy of ploughing and re-seeding with these improved grass seeds—as temporary leys. But the same policy applied to the fine permanent pastures of, say, the western side of England and of the grazing pastures of Leicestershire, Northamptonshire, etc., is nothing less than a tragedy.

One of the dangers of this re-seeding is the very purity of the seed, making such pastures dangerous for the grazing of cattle for years. Cattle get blown on them; and they can only be used for producing crops of hay in the first place, so far as cattle are concerned. Good permanent pastures, properly manured, regularly harrowed and rolled, heavily stocked and rested (an impossibility in these days of reduced pastures) give results which, as all experienced graziers and milk producers know, are amazing. Good permanent pastures have values which cannot be assessed: they contain what are known as weeds—which are herbs, well known to cattle, who select them as required, and which are essential to their health; these are lacking in the new leys—hence the danger to cattle of being blown. I have an Aberystwyth-seeded pasture (an expert classed it, two years ago, as 100 per cent perfect) seven years down. I actually had cows blown on this at the end of last March (these were put in for a few hours one day before it was put up for hay). It produced one and a half tons of hay per acre in June—and the only possible way of grazing it at this age was to put cattle in it directly the hay was carried and to keep it closely grazed all the time. In my opinion it will take another twenty or thirty years before this pasture is the equal of permanent pastures of great age on each side of it.

Last winter my Jersey cows were given the hay off this pasture—followed by the hay off one of my permanent pastures. My herdsman reported a definite increase of milk from the latter: he then fed hay (from a temporary ley) of excellent quality which I bought from a neighbour —followed by some hay bought from another neighbour off a permanent pasture; this was full of thistles, but good sweet hay; the same result was apparent—a marked increase of milk from the latter. The rough-looking permanent river meadow pastures of these parts have feeding values beyond assessment. Both milk and beef can be produced from grass more cheaply, and of far superior quality than from any other foodstuffs. I am getting quite good grazing now off permanent pastures which have been heavily grazed since early April.

In the old farm agreements there was a clause stating that the tenant would have to pay a fine of £50 an acre for ploughing up permanent grass. In the opinion of many experienced farmers our ancestors were wiser men than those responsible for the present policy.

I am, sir, your obedient servant,

O. G. S. CROFT

Hephill, near Hereford.

The above letter and my own observations when I visited Major Croft's farm during the summer of 1944 confirm what I noticed many times in the meadows of La Crau and suggest four things: (1) that the current work on the improvement of grassland in Great Britain should be widened to include the botanical composition of the best meadows and pastures still left to us; (2) that all such future studies should deal with the quality of the produce from the point of view of the grazing animal and of the milk yield; (3) that the efficiency of the mycorrhizal association in our grassland should in all cases be determined, and (4) that as soon as conditions permit it, a detailed study of the celebrated meadows of La Crau, including the composition of the irrigation water, should be undertaken and the results published all over the world.

Future grassland investigations might also include the effect of subsoiling on our permanent pastures, meadows, and temporary leys. There is a mass of evidence which points to a shortage of oxygen in the soil under the turf in most of our grass. This limiting factor can be very effectively removed by a subsoiler drawn by a caterpillar tractor. This matter of subsoiling is dealt with in greater detail later (p. 200). It is mentioned here to reinforce the suggestion that the current work on grassland in Great Britain, valuable and stimulating as it undoubtedly is, might be still more useful if it were more thorough and much more fundamental.

THE MAINTENANCE OF OUR BREEDS OF POULTRY

One other problem in regard to the management of livestock must be briefly examined. Of recent years difficulty in maintaining our breeds of poultry has become acute. As is well known, the concentration of laying hens in batteries, although it may increase the supply of low-quality eggs, is useless for carrying on the line. The problem is how best to maintain the vigour of the breeds.

In the course of my European travels I came across examples of poultry keeping which might solve this problem. It is usual to maintain the vigour and martial spirit of game birds by keeping them out of doors in a wood. The adults and the chicks roost in the trees no matter the weather and this preserves their well-known characteristics intact. If they are kept in buildings, the cocks become "runners" instead of warriors.

If the fox difficulty could be solved, there seems no reason why this outdoor system should not be adopted for our breeding strains. But Mr. Thomas Turney pointed out in a recent paper to the Farmers' Club that we cannot keep poultry out of doors on free range and also preserve our foxes. One or other must give way. A solution might be found by breeding our foxes on some island for the various packs of hounds, releasing the males only when needed for the chase. Any which escaped the hounds could be shot at sight. Another method, which I saw in operation at the Co-operative Wholesale Society's bacon factory at Winsford in Cheshire, would be to house the poultry during the night in the open in suitable fox-proof wire-netting cages.

For the study of disease and its prevention poultry possess many obvious advantages. The life of these birds is a short one, they mature very quickly, their maintenance costs little, and definite results can be obtained in a few months.

IO

SOIL FERTILITY AND HUMAN HEALTH

In the last two chapters the relation between soil fertility and the health of crops and of livestock was discussed. But what of the effect of a fertile soil on human health? How does the produce of an impoverished soil affect the men and women who have to consume it? The purpose of this chapter is to show how an answer to these questions is being obtained.

When discussing how crops and livestock are influenced by an impoverished or by a murdered soil, the subject is obviously restricted to the solid portion of the earth's crust, because cultivated plants and domesticated animals are nourished by what the earth's green carpet produces. But when we consider mankind, we have to include the liquid portion of this planet—oceans, lakes, and rivers—which provide a proportion of our food. We must also take note of the produce of the large area of uncultivated land in the shape of forests, prairies, and so forth, which produce some fraction of our nourishment. These additional sources of food have not been sensibly altered by *homo sapiens*. He has so far not seriously attempted to increase the harvest of the sea by means of chemical manures or to interfere with the natural produce of the forest or the prairie. These have escaped the attention of agricultural science and their crops are now what they have been for centuries— Nature's unspoilt harvest.

We must further include in a survey of our consumption the wholeness of produce as created by Nature. This point is of the greatest importance in considering such things as our daily bread. Freshness is another factor, particularly in vegetable food. Finally we must consider the influence on our general nutrition of the various food preservation processes such as canning, dehydration, and freezing.

The food supply of civilized man is, therefore, a wide subject. Its investigation bristles with difficulties—some are inherent in the subject, others are man made. For all these reasons we must, therefore, not expect to obtain such rapid and such clear-cut results as are easily possible when considering the relation between soil fertility and the health of crops and livestock.

Let us first consider the difficulties which are inherent in the subject. These are at least three. In the first place, the average expectation of life of a human being is many times that of the average crop and of most of our domesticated animals; human beings also carry large reserves which can easily be used. Any results on health due to the food supply are, therefore, likely to develop slowly. In the second place, we cannot experiment on human beings in the same way as we can on crops and animals. Lastly, it is at present almost impossible to obtain regular supplies of the produce of well-farmed land, with which to feed a group of people for the time needed to show how such produce influences their health and well-being. Except in a very few cases, food is not marketed according to the way it is grown. The buyer knows nothing of the way the land was manured—or poisoned. The only way to obtain suitable material would be for the scientific investigator himself to take up a piece of land and grow the food. This, so far as my knowledge goes, has not been done. This omission alone explains the scarcity of reliable experiments and results, and why so little real progress has been made in human nutrition. Most of the laboratory work of the past has been founded on the use of material very indifferently grown. Moreover, no particular care has been taken to see that the food has been eaten fresh from its source. The investigations of the past on which our ideas of nutrition are, for the moment, based have, therefore, little or no solid foundation.

When we come to consider the man-made obstacles that have to be overcome in any investigation of human nutrition, we reach what may fairly be called the citadel—the fortress, as it were, that must first be reduced before the final investigations which are needed can even begin. These difficulties are bound up with the present-day organization of the medical profession. As is well known, our doctors are not only trained to study and cure disease, but receive their remuneration either from the State or from their patients for these duties. The general outlook of our medical men is, therefore, pathological: like any other profession they have to consider how to make a living in return for the services they render: they have also organized themselves somewhat on trade union lines. There is now little or no training for positive health: no openings and no remuneration exist for the pioneer who wishes to ascertain and

174

demonstrate the connection between soil fertility and health. The great prizes of the profession lie in the opposite direction—in surgery and in conventional medicine. There is no Harley Street in which the apostles of real preventive medicine can be found and consulted.

But thanks to the work of the pioneers of the profession itself, a change is taking place. The importance of positive health, of real preventive medicine, and the reform of medical education and training, so that an altogether new type of medical man can be created, fitted to lay the foundation of real preventive medicine—the public health system of to-morrow—are now being actively debated. Naturally these include the whole future of the medical profession, of our hospitals, and the place of the State in the new organization. As there will be no source of private remuneration for men and women engaged in promoting health and preventing disease at its source, the State is the obvious paymaster. The whole movement is a natural development of the present panel system. But the individualists among the medical profession object to their profession coming under the control of the Ministry of this or that. They point out how the dead hand of the permanent government official is certain to stifle all originality, all freedom, and all progress. Judging from my own personal experience of the way the State has ruined agricultural research, there is much to be said for seeing to it that the apostles of preventive medicine must have scope, freedom to work out their own salvation, and above all protection from the petty interference of the average bureaucrat who, at any moment, may be promoted to control men immeasurably superior to himself.

The problem is the age-old one of reconciling the claims of the individual and of the organization. State service pure and simple suggests no solution for such a case: it would merely provide an example of what to avoid. But this does not mean that no solution is possible. The judiciary, for example, is constantly recruited from an endless stream of able lawyers who carry out their work quite independently and unhampered by the Civil Service. No Ministry of Justice exists or is likely ever to be created in Great Britain. Surely the medical profession could regulate a new system of public health very much as the judges manage their affairs. The function of the State and of its various ministries would merely be to provide the funds necessary and then to efface themselves as rapidly and completely as possible. Intimately connected with the creation and regulation of a new system of public health is the reorganization of medical education and training, and the automatic elimination of unsuitable candidates for what amounts to a new profession. Once preventive medicine gets under way fewer and fewer doctors will be needed: the standards for admission will automatically rise. In this way the dictum of Alexis Carrel—the best way of increasing the intelligence of scientists is to reduce their number—can be extended still further. Soon the perfect instrument for the study of health and the reduction of disease will become available. It will be a natural offshoot of the new system.

But what of the intervening period that will be necessary while the new weapon is being forged? We cannot change over suddenly from disease to health; there will be a long time-lag before the old order can yield to new. Two systems must, therefore, exist for a time alongside one another. The old will undergo a natural liquidation; the new will grow from strength to strength.

But while no effective system of public health yet exists, nevertheless there has been progress. The pioneers in the medical profession itself, such as the Cheshire panel doctors and the creators of the Peckham Health Centre, have already blazed the trail. That they have done so speaks volumes for what the profession can and will do in the future. It furnishes the best possible reply to those who say that our doctors think more about money than they do about their work. This I know from long experience and many contacts with medical men all over the world and in many countries to be entirely devoid of truth.

In 1939, in Chapter XII of *An Agricultural Testament*, I summed up the evidence then available for the thesis that soil fertility is the real basis of public health, and in this account dealt with the Medical Testament of the Cheshire doctors and with the work already accomplished at the Peckham Health Centre. The reader is referred to this account and also to Chapter VII of Lady Eve Balfour's *The Living Soil*, first published in 1943, in which further evidence is set out in detail, and to *Pay Dirt,* by J. I. Rodale (Devin-Adair, 1945) which presented *The Medical Testament* of the Cheshire doctors in full, for the first time in the United States. All that is needed now is to emphasize the significance of a few of the older investigations and to describe some of the still more recent results.

Perhaps the most significant of the first set of examples which supported the view that soil fertility is the real basis of public health is that of the people of the Hunza valley to the north of Kashmir in the heart of the Karakoram Mountains with Afghanistan on the west, the Russian

Pamirs on the north, and Chinese Turkestan on the east. Several accounts of the remarkable health of this ancient people have been published based largely on the observations of McCarrison, who at one time was Medical Officer to the Gilgit Agency. In his Mellon lecture delivered at Pittsburgh in 1921 on "Faulty Food in Relation to Gastro-Intestinal Disorder" he referred to the remarkable health of the Hunzas in the following words:

"During the period of my association with these people I never saw a case of asthenic dyspepsia, of gastric or duodenal ulcer, of appendicitis, of mucous collitis, of cancer. . . . Among these people the abdomen over-sensitive to nerve impressions, to fatigue, anxiety, or cold was unknown. Indeed their buoyant abdominal health has, since my return to the West, provided a remarkable contrast with the dyspeptic and colonic lamentations of our highly civilized communities."

The remarkable health of these people is one of the consequences of their agriculture, in which the law of return is scrupulously obeyed. All their vegetable, animal, and human wastes are carefully returned to the soil of the irrigated terraces which produce the grain, fruit, and vegetables which feed them. But there is another replacement in addition to the organic factor. The irrigation water used on these terraced fields comes from the Ultor glacier and is rich in silt. In this way the mineral constituents of the soil are constantly being replaced. How far is the health of these people due to this additional factor? It is impossible to say at the moment. But a growing body of evidence is coming forward in support of the view that to obtain the very best results we must replace simultaneously the organic *and* the mineral portions of the soil. If this should prove to be a general principle, it would help to explain the remarkable health and endurance of many of the hill tribes to the west and north of India where something approaching the Hunza standard is the general rule. In any future investigation of the need for replacing the minerals of the soil Hunzaland is the ideal starting point, as it is a ready-made control station for such studies. Readers interested in this people should begin with "The People of the Hunza Valley," which has just been published as a supplement to No. 9 of the *News-Letter on Compost* (June 1944).

The second of the older examples I should like to comment on relates to the labour force employed by the Public Health Department of Singapore. The results are given in the following letter from the Chief

Health Officer, Dr. J. W. Scharff, to the Editor of the *News-Letter on Compost*. It appeared in the issue of October 1942 (No. 4):

THE SINGAPORE HEALTH DEPARTMENT COOLIES

Rydal Mount,
Potters Bar,
Middlesex.
7th September, 1942.

Dear Dr. Picton,

You have asked me to give you an account of my observations on the health-giving effects of eating freshly grown vegetables grown on soil nourished with compost. The compost to which I refer was made according to the Indore method; an account of how this compost was prepared is published in the *News-Letter on Compost*, No. 2.

From January, 1940, until January, 1942, I had a unique opportunity, due to war-time needs, of watching the progress of a campaign for growing vegetables and seeing that they were eaten by a labour force of nearly 500 Tamil coolies. These men were employed by the Singapore Health Department in various parts of the island of Singapore. As soon as England became involved in war, it became possible to allocate an area totalling in all about forty acres of vegetable allotments on favourable terms to the men engaged on sanitary duties. My labourers were granted these allotments on condition that they prepared compost and used the vegetables and fruit grown therein for themselves and their families only. Sale of the produce was not allowed. Thus it was ensured that these goods were used at home. The local Agricultural Department lent their inspectors and staff to teach the men how best to grow vegetables and demonstrations in cooking and preparation of the foodstuff were organized for each of the labour settlements. Compost making was started on a large scale and during the months previous to the opening of the campaign a supply of over a thousand tons of compost was ready to launch this great experiment.

During the course of the ensuing months apathy and indifference on the part of the labourers gave way to interest and enthusiasm, as soon as it became apparent how well plants would grow on soil rendered fertile with compost. A number of vegetable shows were arranged, at which the healthy produce of fertile soil was exhibited and prizes were awarded. Within six months the accumulated stocks of compost were used up and more active steps were taken to augment the supply, as well as to satisfy the growing demands of other enthusiastic gardeners inspired by the achievements of my men.

At the end of the first year it was obvious that the most potent stimulus to this endeavour was the surprising improvement in stamina and health acquired by those taking part in this cultivation. Debility and sickness had been swept away and my men were capable of, and gladly responded to, the heavier work demanded by the increasing stress of war. But for the onslaught by the Japanese which overwhelmed Malaya, I should have been able to present a statistical record of the benefit resulting from this widespread effort of vegetable culture on compost such as would astonish the scientific world. The results were all the more dramatic in that I had not expected this achievement.

The numbers taking part in this venture were so large as to preclude any possibility of mistake.

It might be argued that the improvement in stamina and health amongst my employees was due to the good effect of unaccustomed exercise or in the increased amount of vegetables consumed. Neither of these explanations would suffice to explain the health benefit amongst the women, children, and dependents of my labourers, who shared in this remarkable improvement. Shortly before the tragic disaster which has brought Singapore within the hateful grasp of the Japanese invader it became apparent that the health of men, women, and children, who had been served consistently with healthy food grown on fertile soil, was outstandingly better than it was amongst those similarly placed, but not enjoying the benefits of such health-yielding produce. An oasis of good health had become established, founded upon a diet of compost-grown food.

This has served me as an inspiration to carry on with this work in whatever part of the world it may now fall to my lot to serve mankind.

Yours sincerely,

J. W. SCHARFF

This interesting nutrition experiment was interrupted by the fall of Singapore. Fortunately Dr. Scharff managed to escape on the last minesweeper which left the fortress and in due course reached England, where he at once resumed his activities on the relation between soil fertility, nutrition, and health—at first in connection with the pig clubs in the London area, and afterwards as a Colonel in the R.A.M.C. at the

179

military camps near Aldershot. He intends, on his return to his old post at Singapore, to continue the work outlined above. His work near Aldershot is being developed with great success by his successor, Major W. H. Giffard.

The value of the above example of the connection between soil fertility and health lies in its simplicity and in the ease with which it can be copied by many employers of labour in the tropics. One such large-scale example in Rhodesia has recently been referred to in the House of Lords by Lord Geddes as follows:

"In 1924 or 1925, when I returned from being on duty in the United States of America for four years, I was asked by the then Prime Minister, Mr. Baldwin as he then was, to see what I could do in connection with the supply of copper for this country. It seems to be a far cry from soil health to copper, but as a matter of fact the nation would not be getting its copper to-day unless somewhere in the back of my mind had been the fact that soil health was what made health. Because the copper that we had to get hold of was in Northern Rhodesia. It was the only place in the sterling area where there were known deposits of copper. It was not very well known, but copper was known to be there because it appeared in native use, and we had to get a copper reserve in order that we might in this country be in a position to defend ourselves, because copper is extraordinarily important in connection with war preparations. The country in which that copper existed was in large parts depopulated. There was no one living there, not even Africans, because of sleeping sickness, malaria and all the range of tropical diseases which make some of the great forest areas in the heart of the tropics impossible for human life. We started in, and the greatest medical problem that I have ever known was the opening up of the Copper Belt in Northern Rhodesia—probably the greatest medical problem of our time.

"There are several branches of medicine. There is curative medicine, which divides itself into research into the nature of diseases, and the other part of curative medicine, the care of sick people; there is preventive medicine, which deals with all the problems of keeping a great community healthy; there is tropical medicine, which is really a spawn of zoology—it is rather making a study of the wild animals that live in the country, even though they are small. And then there is creative medicine, and creative medicine is a thing that very few people know anything about at all. In going into Northern Rhodesia

180

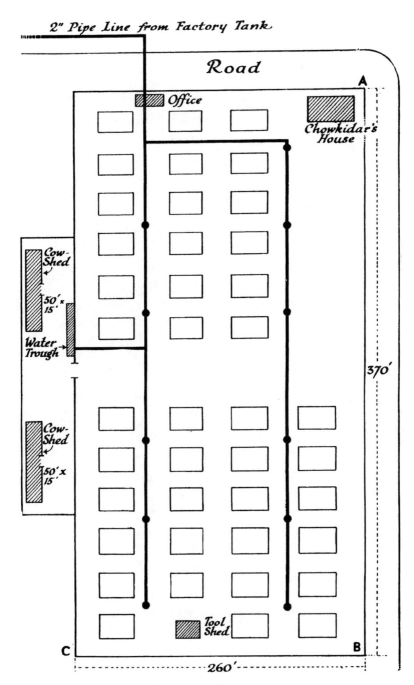

2" Pipe Line from Factory Tank

Road

Office

Chowkidar's House

Cow-Shed

50' x 15'

Water Trough

Cow-Shed

50' x 15'

370'

Tool Shed

C   B

260'

PLATE XI. PLAN OF THE COMPOST FACTORY

GANDRAPARA TEA ESTATE

PLATE XII. COMPOST-MAKING AT CHIPOLI, SOUTHERN RHODESIA

GENERAL VIEW OF COMPOSTING AREA

we had to use all the forms of medicine in order that we could get in. A country that has been depopulated by the virulence of the diseases there is not an easy country to get people of another race into and to keep them in a good state of health. I shall not bore your Lordships with the various steps taken during the fifteen years that followed, but I will tell you this. The curative medicine was just the ordinary sort of curative medicine of Harley Street or elsewhere. It was interesting, but of very much less interest than the other. Preventive medicine dealt with the ordinary problems of public health in a community. As to tropical medicine, the School of Tropical Diseases helped us and we found out a lot of things ourselves. Creative medicine—what did we base that on? On the health of the food; and my noble friend Lord Bledisloe can tell you that our idea of how to keep people healthy there is that we give them food grown on rich humus soil with plenty of life in it.

"What have we done? What have the men who were there done? I do not want to take any credit for myself—I was only chairman of the company. The people who fought the thing through were the doctors and the agriculturists on the spot—everybody there. My job was simply to see that they were not interfered with by short-sighted economy. They have beaten back disease, and turned that part of Northern Rhodesia into what is a health resort. It is a most extraordinary phenomenon. The positive health of these people is based on food. This group of facts provide evidence tending in a definite direction. They show the importance of what Lord Teviot has brought before your Lordships, and they show it in a way, I believe, that places the truth of his contention on a secure basis—that food is the basis of health, but it is not the only basis."

In the same speech Lord Geddes referred to the health of the people of Prince Edward Island in the Gulf of St. Lawrence. This is a relatively small community, made up almost entirely of the descendants of western European stocks (Scots 44 per cent, English 21 per cent, Irish and French 35 per cent). "There we have a very high standard of health, an extraordinarily vigorous, active population and no fall whatever in the birth rate. It is the only social organization composed of western Europeans which has not shown in the last fifty years a really sharp fall in the birth rate." The population is composed of fishermen, farmers, and of craftsmen engaged in rural trades. There are no great cities. The farming is mixed, little artificials are used, and the land is kept fertile

181

by means of muck and the harvest of the sea. (*Parliamentary Debates, House of Lords, Vol. 129, No. 98, 26th October 1943.*)

Details of these two cases—the Rhodesian Copper Mines and Prince Edward Island—have been given in full for two reasons. They are of the greatest interest and value in themselves: they suggest the need for a further detailed description, if possible carried out by an apostle of preventive medicine. If a report could be drawn up on both these examples, the man in the street interested would be provided with definite cases of the way soil fertility and health are influencing one another.

A third little-known example is provided by St. Martin's School, Sidmouth, where for many years the vegetables and fruit needed in the school were raised from fertile soil. The results obtained are summed up in the following letter dated 24th November 1943 from the head master, the Rev. W. S. Airy:

"When I opened a preparatory school at Sidmouth in 1914, I was fortunate in finding a residence equipped with one acre of vegetable garden and another of fruit trees, together with the guidance of a wise and gifted gardener. From the day I came, no type of artificial manure or fertilizer has been known on the premises. Our soil, which has always been dressed annually with some ten or twelve tons of farmyard manure, the contents of poultry houses in the grounds, and two compost heaps, enjoys immunity from insect pests and disease.

"From 1914 to 1941, when war conditions compelled us to give up boarders, all boys were daily supplied with an abandance of fruit and vegetables; also with lettuce, radishes, cucumbers, and tomatoes in season. They were provided with savoys, cauliflowers, beet, onions, peas, beans, parsnips, asparagus, etc., which all flourished in perfect condition. Our exceptional health record has been chiefly due to the school menu. I firmly believe that this would have proved impossible had not the soil been maintained in a superlative state of fertility by means of compost beds and farmyard manure. Epidemics were unknown during the last fifteen years. We had many lads who came to us as weaklings and left hearty and robust; they never looked back in point of health, and are now playing a prominent part in the world crusade of to-day. It has always been my conviction that health, strength, and self-reliance are mainly dependent upon the quality of feeding in preparatory schools at the critical period between nine and fourteen years."

A fourth example is that of St. Columba's College, Rathfarnham, near Dublin, an illustrated account of which was published in the issue of

Sport and Country of 17th March 1944. This is a somewhat complete example. The boys of this college in their spare time are doing a good deal of the manual work of a farm of some 200 acres, fifty acres of which are under cultivation, where most of their food is grown by means of compost made on the spot from animal and vegetable residues. This boy labour is voluntary and supplements that of a paid staff of experienced land workers. Produce is sold by the farm to the college at market rates, and in this way the farm has been able to pay its way. There is no doubt that the experiment has been of immediate practical value in helping to solve the wartime difficulties of catering. The health of the community generally has been unusually good, and the work and games have been continued with additional zest. The current work on the farm and the biological teaching have been made to supplement one another. The medical officer of the college is now preparing an account of this interesting experiment from the health point of view.

A fifth example—of factory canteen meals of the right standard—must be quoted. This has been provided by the Co-operative Wholesale Society's bacon factory at Winsford in Cheshire. These pioneering canteen meals at Winsford are the result of the interest of the manager, Mr. George Wood, in nutritional problems. The factory is a modern one and at the beginning was surrounded by an area of waste land which has been transformed into a model vegetable garden by means of compost made partly from the wastes of the factory. The potatoes and other vegetables needed in the canteen meals are grown on this land. The potatoes are cooked in their skins, and the whole of the tuber is eaten. The area under cultivation is being increased and soon it will be possible to provide all the food needed for the canteen meals from fertile soil. Only whole-wheat bread is provided. Already the health, efficiency, and well-being of the labour force has markedly improved. The output of work has increased; absenteeism has been notably reduced. Here is an example of what can be accomplished for his workers by a manager with vision and enterprise at no cost to the undertaking, as such factory meals pay their way. The workers benefit by excellent meals, far more nutritious and far cheaper than they can obtain elsewhere. The factory benefits by better and more willing work, by the growth of real *esprit de corps,* and by a marked reduction in ill health. Work begins to go with a swing once the food of the workers comes to them fresh from soil in good fettle. Here is a simple method of dealing with industrial fatigue.

A sixth example comes from New Zealand, where the deterioration in the health and physique of the population has followed closely on the heels of soil exploitation. In *The Living Soil* Lady Eve Balfour has dealt with this case in full. The general health status of the population will be clear from the following extract taken from her book (p. 131):

"Of every hundred children who enter New Zealand schools, fifteen show signs of needing medical attention, fifteen need observation, many show signs of nose and throat trouble, and at least two-thirds have dental caries. In this connection, the New Zealand Ministry of Health has published the fact that 30 per cent of all pre-school children suffer from nose and throat troubles, 23 per cent suffer from gland troubles, and 2 per cent have some form of lung trouble. The official figures for illnesses among children at school are: 5 per cent suffering from enlarged glands; 15 per cent suffering from incipient goitre; 15 per cent suffering from enlarged tonsils; 32 per cent suffering from dental caries; and 66 per cent suffering from other physical defects."

At this point Dr. G. B. Chapman comes into this dismal picture. In 1936 he set in motion a feeding experiment at the Mount Albert Grammar School at Auckland. The fruit and vegetables needed by some sixty boys, teachers, and staff were grown on humus-filled soil. The results are reported by the Matron in the following words:

"The first thing to be noted during the twelve months following the change-over to garden produce grown from our humus-treated soil was the declining catarrhal condition among the boys. Catarrh had previously been general and, in some cases, very bad among the boys. In specific cases the elimination was complete. There was also a very marked decline in colds and influenza. Colds are now rare and any cases of influenza very mild. Coming to the 1938 measles epidemic, which was universal in New Zealand, the new boys suffered the more acute form of attack; while the boys who had been at the hostel for a year or more sustained the milder attacks, with a much more rapid convalescence.

"During the past three years there has been a marked physical growth and development during terms of heavy school work and sport. In some cases boys go through a period of indisposition for several weeks after entering the hostel. This would appear to indicate that the method of feeding causes a certain detoxication period, which, when cleared up, does not return. Excellent health gradually ensues in all cases, and is maintained. There are fewer accidents, particularly in the football

184

season, which would possibly indicate that the foods in use contain the optimum amount of minerals and vitamins, thus ensuring a full development of bone and muscle and a greater resiliency to fracture and sprains. The satisfactory physical condition described is maintained during periods of rapid growth and the development of mind and body. Constipation and bilious attacks are rare. Skins are clear and healthy, while the boys are unceasingly active and virile.

"Since the change to naturally grown garden produce, the periodical reports in regard to the boys' dental condition have been more than gratifying."

The deterioration in the general health and well-being of the New Zealanders and the above timely intervention on the part of Dr. Chapman have been followed by a most interesting and promising development in the shape of a Compost Club, details of which are given in a later chapter (p. 237).

After this book had gone to press a significant report reached me from Mr. Brodie Carpenter, the dentist in charge of the teeth of some 97 girls and 137 boys at a boarding school in Middlesex, where during the global war great attention had been paid to the growing of the vegetables and salads on humus-filled soil without any help from artificial manures. A full report on the methods adopted in the raising of this produce, of the composition of the school meals and their effect on the teeth of the children appeared in the issue of *The News-Letter on Compost* of February 1945, pp. 21–2. In 1939 when the experiment started the standard of the teeth was distinctly poor. By 1942 a change began to take place: by May 1944 a vast improvement had occurred—the general standard had gone up to good.

What is needed to bring home to the man in the street the supreme importance of soil fertility as the basis of the public health system of to-morrow are more and more examples of what a fertile soil can do. The type of examples needed will be clear from those already quoted. *Boarding schools and colleges should produce at least their own vegetables and fruit from humus-filled soil.* The labour difficulty will disappear the moment the teaching staff, the boys, the girls, and the students understand the importance of the question. This is proved by the example of St. Columba's College—the Eton of Eire. The school gardens and canteen meals of our elementary schools can easily copy what has already been done at the Mount Albert School in New Zealand. Full details of the best way to grow and to cook vegetables raised in a school garden are to be found in Mr. F. C. King's book, *The Compost Gardener* (Titus Wilson & Son Ltd., Kendal). Factory canteen meals might with advantage copy what Mr. Wood has done at the bacon factory at Winsford in Cheshire.

But perhaps what would be the most telling example remains to be discussed. *That seaside holiday resort which takes steps to have produced from fertile soil in the neighbourhood most of the food needed by the visitors would rapidly forge ahead and out-distance all competitors.* Holiday-makers need rest, good air, and above all good food. If an autonomous community like the Isle of Man could become compost-minded and see to it that most of the food needed by the visitors was grown locally on fertile soil, it would rapidly become the most popular holiday resort in Great Britain. Steps could then be taken to provide the stream of satisfied visitors with details of how to get their own gardens and allotments into shape, so that the good work started in the Isle of Man could be continued till the time for the next seaside holiday came round.

THE NATURE OF DISEASE

I<small>N THE FOUR</small> preceding chapters the diseases of the soil, the crop, the animal, and mankind have been discussed, and my observations and reflections on these matters have been recorded. This recital is of necessity somewhat fragmentary, because such a mass of apparently unrelated detail has had to be described. At least one question will occur to the reader at this point: Is there any underlying cause for all this disease? If the birthright of every plant, animal, and human being is health, surely all these examples of disease must have something in common. It has been suggested throughout these chapters that much of this disease is due to farming and gardening methods which are inadmissible. If this is so, how do these mistakes in practice operate?

For many years I have been on the look-out for some guiding principle which would explain matters and feel convinced that I have at last found it in the writings of a distinguished investigator of human diseases—Mr. J. E. R. McDonagh, whose work is not very widely known, due perhaps to the fact that an attempt has been made by the author to convey a too complete scientific picture of a very difficult and very intricate subject. I have therefore asked Mr. McDonagh to set out in the simplest possible language the gist of his results on the nature and causation of disease which are discussed in full in his *The Universe Through Medicine* and other writings. He has very kindly done so in the following note dated 8th September 1944:

"*The Nature of Disease.* Every body in the universe is a condensation product of activity. Every body pulsates, that is to say it undergoes alternate expansion and contraction. The rhythm is actuated by climate. Protein in the sap of plants and in the blood of animals is such a body, and it is also the matrix of the structures in the former, and of the organs and tissues in the latter. If the sap in plants does not obtain from the soil the quality nourishment it requires, the protein over-expands. This over-expansion renders the action of climate an invader, that is to say climate, instead of regulating the pulsation, adds to the expansion.

The over-expansion results in a portion of the protein being broken off, and this broken-off piece is a virus. The virus, therefore, is formed within, and does not come from without, but protein damaged in one plant can carry on the damage if conveyed to other plants. The protein in the blood of animals and man suffers the same damage if it fails to obtain the quality food it needs. In animals and man a third factor enters, and that is an invasive activity of the micro-organisms resident in the intestinal tract. This activity causes still further expansion, and the tissue and organ damaged is the one which originates from that part of the protein which is made to undergo the abnormal chemico-physical change, hence there is naturally only one disease, and this is regulated by the damage suffered by the protein wherein the host's resistance lies. As a result of the micro-organisms in the intestinal tract having played an invasive role for so long, they have in addition given rise to micro-organisms which can invade from without, but from these few remarks you will see that micro-organisms do not play the causative role in disease with which they are usually credited."

According to this view of disease, the heart of the subject must reside in the proteins. If these are properly synthesized in the plant, their disease-resisting powers first protect the crop and are afterwards duly handed on to the animal and to man. If, therefore, we see to it in our farming and gardening that the effective circulation of protein from soil to plant, and then to livestock and mankind is maintained, we shall prevent most of the departures from health—that is to say, disease—except those due to accidents or to abnormal climatic conditions.

Extremes of climate, by tending to damage the proteins, remain as factors in the causation of disease. We cannot always completely control the climate. For this reason it will be impossible to prevent all disease. We can only reduce its amount and soften, as it were, its incidence.

But in one important direction we can do much to control climate—in the effective regulation of the pore spaces of the soil—where those portions of the plant occur which are least protected—the root hairs and absorbing areas of the root. By maintaining the water and air supplies of these internal portions of the soil—the pore spaces—and also by providing the soil population there with constant supplies of humus of the best quality, we can do much to give this important section of the machinery of our crops ideal climatic conditions. Both the root hairs and the mycorrhizal association can then function effectively. The soil population will also thrive. There will be abundant material for repairing the compound particles: so soil erosion will become impossible. The microbial life of the soil will remain aerobic, so the formation of alkali soils will not occur.

In the case of livestock and mankind the extremes of climate can, of course, be mitigated by the provision of fresh food from fertile soil and by providing warmth and shelter. All this will help the proteins to carry out their duties in resisting the onslaught of all kinds of invaders and in the prevention of virus diseases.

The synthesis of proteins in Nature is intimately bound up with the nitrogen cycle. The proteins made in the green leaf represent the last phase in this nitrogen cycle between soil and plant. When these proteins are manufactured from freshly prepared humus and its derivatives, all goes well; the plant resists disease and the variety is, to all intents and purposes, eternal. But the moment we introduce a substitute phase in the nitrogen cycle by means of artificial manures like sulphate of ammonia, trouble begins which invariably ends with some outbreak of disease and by the running out of the variety.

A simple explanation of the relation of soil fertility to health is thus provided. All my own experiences and observations fall into line with this principle. The cure, by growing the affected plants in freshly prepared compost, of virus troubles in crops like strawberries, raspberries, tobacco, and sugar-cane, is explained. Imperfectly synthesized protein is then replaced by normal protein.

In all future studies of disease we must, therefore, always begin with the soil. This must be got into good heart first of all and then the reaction of the soil, the plant, animal, and man observed. Many diseases will then automatically disappear. Only the residue will provide the raw material for the studies of the diseases of to-morrow.

Soil fertility is the basis of the public health system of the future and of the efficiency of our greatest possession—ourselves.

How the vast amount of humus needed to get the soil of the agricultural world into real shape can be prepared and used will be dealt with in the third section of this book.

189

PART III

THE PROBLEM OF FERTILIZING

12

ORIGINS AND SCOPE OF THE PROBLEM

THE GREAT PROBLEM before agriculture the world over is how best to maintain in health and efficiency the huge human population which has resulted from the Industrial Revolution. As has already been pointed out, this development is based on the transfer of food from the regions which produce it to the manufacturing centres which consume it and which make no attempt to return their wastes to the land. This amounts to a perpetual subsidy paid by agriculture to industry and has resulted in the impoverishment of large areas of the earth's surface. A form of unconscious banditry has been in operation: the property of generations to come, in the shape of soil fertility, has been used not to benefit the human race as a whole, but to enrich a dishonest present. Such a system cannot last: the career of the prodigal must come to an end: *a new civilization will have to be created*, in which the various reserves in the earth's crust are regarded as a sacred trust and *the food needed is obtained* not by depleting the soil's capital, but *by increasing the efficiency of the earth's green carpet*. This involves the solution of the problem of fertilizing.

Why does the problem of fertilizing arise? What is the reason for our constant anxiety about the state of the soil? This preoccupation is as old as the art of agriculture. The problem occurs throughout the world, being recognized as a first consideration among all cultivating peoples. Its antiquity and its universal character are striking and must lead us to conclude that it is based on something of fundamental importance.

Briefly stated, the necessity for manuring arises out of our interference with the natural cycle of fertility. It is perhaps the most insistent of those problems which owe their origin to human action directed towards manipulating for the benefit of humanity the life of the vegetable and animal kingdoms. For be it admitted, the operations of cultivation, sowing, and reaping—all the acts that make up agriculture—are serious interruptions or interventions in the slow and intricate processes which make up growth and decay.

This is, perhaps, the place to devote a few words to this basic conception of agriculture as an interference with Nature. I have been attacked for not recognizing that interference. My constant references to Nature as the supreme farmer have been found inapplicable and inept, it being pointed out that if we were to follow Nature alone, we should be restricted to those small harvests which she is accustomed to provide, to the gatherings from the woodland and the hedgerow, from the wild pasture or the moor. I am accused of ignoring the fact that the whole aim of the cultivator is to do better than Nature and that the success attained in this direction is a source of legitimate pride.

It is, therefore, not out of place to take this opportunity of stating that the conception of agriculture as an interruption or interception of natural processes has always been recognized by me.[1] Where I part company with my critics is in my general view of the unbalanced nature of these human acts. Intervention there must be: the most elementary act of harvesting is an interception: the acts of cultivation, sowing, and so forth are even more deliberate intrusions into the natural cycle. But these interruptions or intrusions must not be confined to mere exploitation: they involve *definite duties to the land which are best summed up in the law of return:* they must also realize the significance of the stupendous reserves on which the natural machine works and which must be faithfully maintained.

The first duty of the agriculturist must always be to understand that he is a part of Nature and cannot escape from his environment. He must therefore obey Nature's rules. Whatever intrusions he makes must be, so to say, in the spirit of these rules; they must on no account flout the underlying principles of natural law nor be in outrageous contradiction to the processes of Nature. To take a modern instance, the attempt to raise natural earth-borne crops on an exclusive diet of water and mineral dope—the so-called science of hydroponics—is science gone mad: it is an absurdity which has nothing in common with the ancient art of cultivation. I should be surprised if the equally unnatural modern practise of the artificial insemination of animals were not also to be condemned. Time will show.

But, provided that the actions of the cultivator are well conceived,

---

[1] See especially what I wrote for students in a small book on *Indian Agriculture*, p. 11, which was published by the Oxford University Press in 1927.

that they have been proved successful by long experience, that they follow the essential course of Nature without real disobedience, that *the character of the intervention undertaken is comprehended and that measures are initiated to restore the natural cycle in a proper way,* much may be accomplished by man: and *this* is the art of agriculture.

The final proviso is of the utmost importance; we must give back where we take out; we must restore what we have seized; if we have stopped the Wheel of Life for a moment, we must set it spinning again.

Such a conception is very different from the all too prevalent idea which sees Nature as a parsimonious and very sparing provider of scanty, dispersed, and irregular harvests, a force which has to be stimulated by chemicals into adequate response, and controlled by the ingenuity and inventions of modern times. On this ingenuity and on those inventions rests, so it is claimed, the constantly growing food supply needed by modern populations, and much time is devoted to reckoning up the magnitude of this human achievement. The argument is based on figures of increased crop and animal production over the last few generations of human life and ignores the fact that *these results depend on the plunder of the capital of the soil.* The conclusions reached are fundamentally erroneous and are fraught with the certainty of failure and catastrophe.

This want of perspective and lack of humility dominates most of the short-term solutions of the problem of fertilizing, which from its very nature calls for the closest consideration of natural law. Without further ado I therefore propose to return to my usual method of first reflecting on the natural processes governing the question at issue, then examining what departures from these processes have been made by human action, and finally asking my readers for a sympathetic consideration of a certain point of view which may in some respects be new and even surprising.

The methods adopted by Nature for maintaining the earth's surface in fertility have been referred to throughout this book. They need only be briefly summed up here.

There is first a slow creation and interchange of soils by means of weathering and denudation through the agency of water or wind. Soils are constantly being shifted and redistributed. This long, slow process prevents the earth's soils from becoming static, in fact from becoming stale and worn out: we have only to imagine what would be the state of affairs as regards the supply of minerals if this process of natural regeneration did not take place. Secondly, *there is a vertical movement whereby the roots of trees draw up the minerals of the subsoil, which then become distributed by the leaf fall.* The constituents of the subsoil are thereby, and by means of the earthworm, continually being added to the top soil. There is thirdly the deposit on the surface of new organic residues everywhere on a colossal scale: these are derived from all vegetable growths—trees, grass, or whatever they may be—which are agents for catching and using the power of the sun, the final source of fertility. Fourthly, there are animal wastes, both the wastes from living creatures and the decomposition products of their dead bodies; these wastes in all their forms are in nature always widely dispersed. Finally, these factors of fertility are acted upon, one might almost say directed, by moisture and by air: they are first mechanically mixed and then transformed in their biological, physical, and chemical characters by the action of the smaller animals and invertebrates and by the agency of millions of microscopic fungi and bacteria.

Much of our interference with this complex of process is unavoidable. The settlement of areas for cultivation is a first necessity: we cannot afford to have our farms moved hither and thither. The allocation of chosen crops for selected fields then follows. This is a very violent interference with natural life, which mixes and rarely selects. The consequences of this major interference are made good by systems of rotation and mixed crops, which are designed to restore that variety of vegetable growths which had to be sacrificed for purposes of convenient cultivation: the old device of fallowing is part of the rotation principle. That this restoration of fertility is often very imperfect has already been shown in the chapter on "The Maintenance of Soil Fertility in Great Britain" (p. 43).

Apart from these long-term intrusions there are, especially in Western agriculture and in a great deal of plantation agriculture, short-term omissions—annual, seasonal, and indeed daily—to maintain the fertility cycle. These omissions are mostly unconscious and are, therefore, not being made good by counter-measures: herein lies their danger. There is, first, the general neglect of vegetable wastes: these are not faithfully returned to the fields as they should be: they are sometimes burnt, and they are partly removed for industrial and other purposes and then buried for decades in sealed tips of urban refuse. Far more injurious is the neglect of animal wastes. Human wastes are washed away, while the wastes of domestic animals, often insufficient in volume, are concen-

196

trated in rank manure heaps *instead of being dispersed*. This matter of the dispersal of animal wastes is important.

The effect of these interferences with natural law accumulate and the discussion of the problem might be prolonged on these lines. But the reader has already been put in possession of the gist of the subject; in order not to deflect his attention the remainder of this section of the book will be devoted to special points which seem at the present stage to throw the most light on the vital problem of fertilizing.

THE PHOSPHATE PROBLEM AND ITS SOLUTION

The problem of manuring does not concern the top soil only: it includes the subsoil. The circulation of minerals between soil and subsoil is an essential factor in any manurial programme.

As already stated, the past history of our fields has constituted one of those major intrusions into the natural fertility cycle of which the results are now becoming apparent. Most of these fields were originally under forest. This forest cover would soon be re-created if our arable or pasture land were enclosed and left to itself. This is Nature's time-honoured method of restoring soil fertility. The trees and undergrowth soon accumulate the essential stores of humus; the roots break up the subsoil in all directions and comb it thoroughly for minerals like phosphates, potash, and the various trace elements, which are then converted into the organic phase in the leaves and afterwards transformed into humus for feeding the soil population. At the same time, the roots leave behind them not only a pulverized subsoil, but also numerous channels for air and water, as well as a supply of organic matter. In this way the roots improve the condition of the subsoil; permeability is restored; and, what is equally important, the natural circulation of minerals between subsoil and soil is renewed. Everyone knows how fertile are the soils left by the forest. One reason is that they are rarely short of minerals. The ultimate source of minerals such as phosphates is the primary or igneous rocks, many of which contain appreciable quantities of phosphate in the form of apatite. From these primary rocks the sedimentary rocks are derived. Both classes give rise to subsoils and soils, so that when we look at the phosphate and indeed the mineral question as a whole and start our studies at the source, we should expect any shortages of phosphate or other minerals to be due to some error in

197

soil management. This is exactly what has happened, In the course of years of cultivation the circulation of minerals between subsoil and soil has deteriorated. The constant treading of animals, the passage of cultivating machines, the failure to use afforestation to renew soil fertility,[1] the failure to replace the root system of the trees by those of deep-rooting plants while the land is rested under grass, and the excessive use of chemicals have caused the subsoil to form a definite pan which restricts the passage of roots, interferes with the aeration of the lower layers, and leads to a poor circulation of minerals between the surface soil and the great reservoir of the subsoil. Crops have in this way been forced to live more and more on the thin upper layer of cultivated soil and so have exhausted such elements as phosphorus, potassium, and the trace elements. The soil, therefore, suffers very much as an animal does when the circulation of the blood is defective. *The first matter to attend to, therefore, is to restore the natural circulation of phosphate and other minerals between subsoil and soil.* At the same time we set in motion, through the operations of weathering and denudation, the natural replenishment—from the underlying rocks—of the minerals removed by crops and livestock.

Some fifty years ago during my student days spectacular results were beginning to be obtained when heavy land under grass was dressed with finely pulverized basic slag. Basic slag is the name given to the used-up limestone lining of the Bessemer converter, by which the phosphorus from certain types of iron ore is removed. The molten metal gives up its phosphorus to the limestone with the formation of one of the phosphates of calcium. This, when finely powdered, acts as a phosphatic manure. In this way a new artificial manure was added to an already long list.

The obvious effect of this slag on a piece of heavy land under grass is to improve the herbage, the clovers in particular. But when basic slag is added to pastures on light, permeable land and to grass on the chalk, negative results are often obtained. I well remember how all this

1 In all future afforestation schemes care should be taken to use the forest to improve the areas under agricultural crops. This can most easily be done (1) by starting the new plantations on land which has been subsoiled and brought into good condition by suitable cultivation, temporary leys and by abundant humus, (2) by raising the young trees in humus-filled nurseries so that the mycorrhizal association can be established from the beginning, and (3) by a suitable mixture of trees. In this way the time taken to grow marketable timber could be vastly reduced and the income of the new plantations increased. As soon as possible these afforested areas should be cleared and then given back to agriculture. Another area could then be put under this long term forestry rotation. In this way forestry can be used to restore the fertility of the soil as well as to provide timber. The marriage of forestry and farming must be included in all our future agricultural policies.

troubled me when I connected these results with my knowledge of geology and of the microscopic structure of the primary rocks. Something seemed to be wrong somewhere. I put my doubts to my instructors and suggested that the whole phosphate question should be reopened. Their explanations failed to satisfy me. Then about 1904 at the Royal Agricultural Show at Park Royal a chance observation led, some forty years later, to the practical solution of the phosphate problem. Some turves taken from the plots of the Cockle Park experiments were included in one of the exhibits dealing with agricultural research. One of these turves was taken from the plot which had received basic slag, the one alongside from the control plot. The difference in the herbage was amazing, but what also interested me was the deep, black layer of humus under the slagged turf and the absence of a similar humus layer in the control. Thirty-four years later, in 1938, I was able to continue this phosphate story. I discussed my observations with the late Sir Bernard Greenwell and suggested that basic slag must act indirectly *by improving the aeration of heavy soils,* whereby the vegetable and animal wastes are converted into humus, which in turn would improve the grasses and clovers. I pointed out that under the turf of heavy, close grassland nitrates were always in defect and that the provision of more oxygen invariably improved matters. He at once proceeded to use a subsoiler, drawn by a caterpillar tractor, four feet apart and twelve to fourteen inches deep, on his grassland on the London clay and immediately obtained results comparable with those obtained by an average dressing of slag. The passage of the shoe of this machine acted like a mild explosive and shattered the subsoil. The land, of course, must be in the right condition to obtain the maximum effect—it must not be too wet or the pan will not shatter.

Sir Bernard's death in 1939 and the war program put an end to the work in progress at Marden Park. The results thus so far obtained, however, were set out in 1940 in Chapter VII of *An Agricultural Testament* in the hope that some pioneer would be sufficiently interested to continue this phosphate inquiry. In 1943 the expected happened. I received a letter from a correspondent in Sussex—Mr. R. Delgado, Little Oreham, near Henfield—to the effect that he had prevailed on his local War Agricultural Executive Committee to subsoil one of his pastures. At the same time a neighbour applied ten hundredweight of basic slag to each acre of similar land. In view of the importance of this work, the correspondence is here quoted *in extenso.*

The first report is dated 27th November 1943:

"After reading Sir Albert Howard's *Agricultural Testament* and the account he gives in it relating to the work of the late Sir Bernard Greenwell with the use of a subsoiler on pastures overlying clay, it was decided at once to contact the local W.A.E.C. with a view to finding out the name of a contractor who possessed the necessary tackle to carry out such work on my farm.

"The local Committee wrote back to say that I was misinformed and that the only use a subsoiler had was on arable ground behind a plough. After a further exchange of letters they agreed to send a crawler tractor and a wheel-type Ransomes subsoiler.

"A further argument ensued as to the depth and distance apart, but, after the subsoiler had been up and down the field once, I pointed out to the officer that no effective shattering of the subsoil could take place further than two feet on either side of the share, and he eventually came down to doing them six feet apart and fifteen inches deep.

"Had the work been carried out strictly in accordance with Sir Bernard's stipulation, I am certain that the eventual results would have been better. However, the response from the worst field on this farm was encouraging. When the work was completed, it was stocked with yearling and bulling heifers and three horses. There was not much grass on the field to start with, so good and bad hay were fed to supplement the grazing. The good hay was, naturally consumed and the bad was dunged and trodden on to form compost *in situ*. The field was finally shut up in July absolutely bare, four months after subsoiling.

"Despite the absence of rain in this part of the country during the summer, the flora on this meadow had changed to an emerald green on shutting up and has remained so ever since.

"On November 20th I went to look at it and was agreeably surprised to observe many worm casts which had hitherto been absent. The milking herd was turned into it the following day. Their relish for the short bite was very noticeable, particularly where the worm casts were more numerous, and the milk yield went up.

"In the autumn of last year a friend, farming nearby on the same type of soil, dressed a meadow with ten hundredweight of slag to the acre. I was privileged to see it this June, closely grazed and a very good colour. Everyone is aware of the virtues of slag on clay soils.

"In July, just before shutting up the field described above, my friend paid me a visit and we were standing in this field having a look at my young stock when she remarked on the greenness of my turf, complaining sorrowfully that her slagged meadow was brown, scorched, and devoid of any feed.

"It would be as well to state here that the flora in my particular meadow was that which is found in pastures which tumbled down to grass after the last war with a proportion of volunteer clover and a semi-swamp variety of weeds, whereas in my friend's I had seen a preconceived mixture of grasses and clovers. And in order to complete the treatment of my meadow, after a pulse crop has been taken, it will be sown down to deep rooters and *Leguminosae*.

"There remains the cost of the work. Subsoiling by contract with the W.A.E.C. under the mole-draining scheme, inclusive of piped outfalls averaging three to each four acres and inclusive of a 50 per cent grant, came to a little over 25s. per acre. Subsoiling, as recommended by Sir Bernard Greenwell with one's own power and tackle, one could probably carry out to-day for a maximum of 5s. per acre.

"The cost of slag, which is either £6 per ton or £3 10s., I am not sure which, would work out on a dressing of ten hundredweight to the acre at the lowest at 35s. per acre exclusive of labour.

"Though admitting that slag is better than nothing in that humus formation under the turf sets in after a suitable application, apart from the relative merit of costs I am of the opinion that one can obviate any unknown chemical reaction in the soil by seeking the same, if not possibly better, results by the use of the subsoiler. Unfortunately I have not as yet been able to go and see the slagged meadow this autumn to discover what verdict is given by the earthworm.

"It might be of interest to add that fungi in the shape of mushrooms, only very sparsely scattered in my meadow last year, abounded in great numbers this autumn, whereas it is well known that slag will do away with them for evermore."

The matter was followed up further and in a subsequent report dated 3rd April 1944 Mr. Delgado continues the story of his interesting experiment:

"In April 1943 I subsoiled a four-acre meadow which was literally soused with a century or more of organic decay for about four inches under the turf. I should imagine it was very acid since it hardly grew any hay and the stock loathed it. It was, in fact, one of those meadows which give spectacular results with a heavy dressing of slag. In the previous autumn stock had been shut up in it and fed with green stuff carted from another field.

"Soon after subsoiling, the meadow was ploughed and one-third of it dressed with ten hundredweight of slag to the acre. It was sown with oats and tares. *The crop was uniform throughout the field.* In the autumn of 1943 the field was ploughed again. *The ploughman, who did not know I had slagged a portion of the field, noticed the land was harder on the slagged area.* Winter oats were drilled and at the time of writing (3rd April 1944) the crop is uniform.

"The oats are going to be undersown with a grass mixture, and it will be interesting to see if there is any difference in the take of the seeds as phosphates are supposed to be essential when laying down to grass."

In a further letter dated 22nd April 1944 Mr. Delgado stated that as the oats were very forward he had been compelled to graze them by cattle. The stock grazed the oats evenly and showed no preference whatsoever for the slagged portion. He will continue to keep this field under careful observation and report if any differences develop, and also take note of the reaction of the grazing animal to the following grass crop.

On 20th October 1944, Mr. Delgado reported that the oat crop was uniform and yielded about thirty hundredweight of grain to the acre. The take of the clovers and grasses in the seeds mixture was absolutely uniform all over the field which was evenly grazed by the livestock. As far as could be seen up to the time of writing the application of slag at the rate of ten hundredweight to the acre to a portion of this subsoiled field produced no result.

There seems no doubt that the effect of basic slag is mainly to promote the formation of humus under the turf of heavy land under grass *by improved aeration* and that similar results can be obtained at much less cost by means of the subsoiler.[1]

Mr. Friend Sykes has obtained equally striking subsoiling results on arable land. This he has done by breaking up the pan under the plough sole. His experiences are described in detail in Appendix D to this book (p. 289).

Clearly when war-time scarcity is over and a supply of implements becomes available a regular subsoiling campaign will have to be set in motion throughout the length and breadth of Great Britain. Indeed, in most

[1] "Is Basic Slag Really Necessary?", *News-Letter on Compost*, Nos. 8 and 9, February and June 1944.

parts of the world, systematic subsoiling is certain to be one of the great advances in agriculture. Captain Moubray has already obtained good results in the Mazoe valley in Southern Rhodesia. Some striking effects of subsoiling have also been obtained on the Roosevelt Hyde Park farm in the United States of America. Subsoiling is certain to prove the first great step in maintaining the mineral supplies of the surface soil and so rendering obsolete many of our ideas on manuring. It sweeps current advice on phosphate manuring into the lumber room of exploded ideas. It may also prove to be of great value in the reclamation of alkali land.

Not only does subsoiling open the door to the reform of arable farming, but it will, above all, be a practical solution of some of the problems of our temporary and permanent grassland. Without realizing it, we have in the course of long processes of cultivation allowed our fields and pastures to become pot-bound: *this condition puts at least half of the fertility cycle out of action.* By correcting this condition and allowing air to penetrate beneath the surface down to and into the subsoil, we restore *that natural supply of oxygen without which humus formation cannot properly proceed.* Subsoiling, in fact, is the parallel process to drainage and perhaps, because so long neglected, is even more important: the one process controls the surplus water of the soil and the other guides and restores the supply of air. The soil like the compost heap needs both air and water at the same time.

In this way only can we make a full use of the earth's green carpet, and *it is only by the agency of the green carpet that we are able to trap the sunlight:* in proportion as this green carpet is not utilized we lose that much solar energy. The practical effects of the change are indicated in the reports quoted above. *It is certain that by this reform carried out all over the country the stock-carrying capacity of our grass areas will go up by leaps and bounds.* The door will then be opened to making full use of the improved varieties of grasses, clovers, and herbs—which must always include deep-rooting types and which must also have ample leaf area for intercepting the sunlight—needed by the ruminant stomach. We shall also be able to take in hand all our hitherto neglected second and third classes of land. Most of these will go up at least a class after they have been treated by methods similar to those which Mr. Sykes and Mr. Delgado have so successfully applied at Chantry and at Little Oreham. *The great openings are certain to lie in these and even in fourth-rate areas.* We have only just begun to deal with the hill farms—those cradles of the breeds of livestock of to-morrow. *England need no longer restrict her real farming to the best land as she is doing now.*

THE REFORM OF THE MANURE HEAP

Subsoiling will solve the mineral side of manuring. The reform of the manure heap and the full use of sheet-composting are the roads by which the nitrogen problem must be approached.

If the soil is a living thing, as we have continually been insisting in this book, so also in an even more intense way is the manure heap. Such a manure as compost is simply a teeming mass of microbial and fungous life. This life, like all life, never stands still; it has its own cycles and is in a very different state at different times.

All cultivators like their farmyard manure well rotted. A hot manure, i.e. a fresh manure, cannot safely be introduced into a worn-out soil which is then to grow a crop. This universally accepted piece of practice is a first recognition of the potentially dangerous nature of the traditional heap of farmyard manure—evil-looking, evil-smelling, full of maggots, and the paradise of breeding flies. Our extraordinary habit of heaping up animal excrement together in these insanitary masses is, it is true, established among us by age-old tradition. That must not prevent us from probing into the practice and questioning it.

It is not natural. Nature does not collect the excrement of her fauna in this way. Their droppings in a wild pasture are most widely scattered by the roaming habits of the animals, far more widely than they are even in a field grazed by domesticated beasts. The admitted distaste of such grazing animals for feeding off patches of grass which have been stained, as it is called, by their own wastes some time previously should alone have given us a hint. Horses, for instance, are most particular and may be classed as most cleanly beasts. Nowhere in Nature (if we except a few sea-bird habitats where suitable nesting areas are restricted) do we find the noisome nuisance of the manure heap.

The fact is that by collecting farmyard manure in this way and leaving it, sometimes for many months, at least three deleterious processes are induced.

In the first place, the rain washes out an untold portion of the valuable elements: this is finally lost to the farmer. Whoever has seen the richest part of a large manure heap leaching away into a ditch without hope of recovery may well ask himself why the farmer was at so much trouble to gather together what he is so eager to lose again. *The rich exudation, which leaves the heap, is like an opened artery: all goodness drains away:* a less valuable mass of stuff is left, impoverished of much of the best constituents. Yet this sort of carelessness is met with in almost every farming community outside China, and what is much worse is looked on as nothing in the least abnormal.

In the second place, there is a considerable loss of nitrogen to the air due to the establishment of an anaerobic flora. Though not so obvious as leaching by rain, yet much loss of the valuable element—combined nitrogen—occurs. Such losses are a foregone conclusion if we remember that, as we pointed out above, farmyard manure is not a static substance. Its very nature implies change, just because it is alive. The natural changes it would undergo if left alone would be to become humus by incorporation after fermentation with ample vegetable wastes. But, if not thus left to its natural destiny, if heaped up into a huge solid mound by man's agency, it does not on that account wholly cease to live: and among the living changes which it is bound to undergo is the release of the excess nitrogen by denitrification so that a mixture suitable for humus formation remains. The combined nitrogen it contains, which is so valuable a plant food element and for which the surrounding vegetation is crying out, escapes into the air either in the form of ammonia— the characteristic smell of which hovers over every manure heap—or as free nitrogen gas.

In the third place, something far worse than leaching and the escape of nitrogen is apt to take place in the manure as a final result of cutting off the air supply. Decay in the forms which we have been investigating is one of the ways in which Nature turns her Wheel. It is not, however, her only or exclusive process. There are processes, commonly known as the putrefactive processes, which she also employs in certain circumstances. These processes are always induced when there is insufficient oxygen. In the absence of *oxygen—the great purifying agent which by combining burns up the elements present in decaying bodies*—these putrefactive processes form a special type of compound usually accompanied by the generation of noxious gases. This is putrefaction and we all know, by common experience, what that word means. It is Nature's method of removing wastes which for some reason she is unable to deal with normally by what we may call her methods of healthy decay. Perhaps because there is some stoppage, some kink, in her normal processes, she carries out these alternative putrefactive changes in an unpleasant and sensational way. The sights and smells of putrefaction are highly disagreeable to the higher living creatures, man not excluded. If we like to use a poetical image, it is Nature thwarted, and in wrath.

Now in a manure heap these putrefactive processes are apt to take the place of the normal decay processes, especially when manure is heaped on a concrete floor or in a concreted pit. Any farmer who wishes to observe these putrefactive processes can easily do so by assembling two manure heaps side by side, one on freshly broken-up earth, the other on a concrete floor. The air supply of these two heaps is very different. The first obtains a fair supply of oxygen: in the second aeration is restricted and putrefactive changes, accompanied by an offensive odour, soon set in. Incidentally this simple experiment establishes the principle that the earth itself breathes provided the surface soil is kept open. This is one of the reasons why we must always cultivate.

If putrefactive processes have begun, then the manure is not at a stage suitable for plant food. It will have to undergo some very prolonged changes before the plant can get much benefit from it. Whereas *decomposition without putrefaction* is the principle of compost-making, putrefaction delaying and complicating the normal absorption of food needed by soil and plant is what often follows from the nuisance of the manure heap. The reason is simple. The mere mechanical heaping up of the animal excrement into one large mould has deprived that excrement, first, of the oxygen it needs for burning up, and second, of that juxtaposition and mingling with sufficient waste vegetation of the soil which goes to make normal decay. *We have produced the conditions needed by an anaerobic flora.* This always means loss. We have not mixed the vegetable and animal wastes in the proportions Nature has ordained.

We thus always return to the same point: animal and vegetable *must be mixed* in correct proportions in their death, as in their life, processes.

This criticism of a very ancient practice in agriculture will appear bold. The manure heap has been used by generations of farmers. If there were nothing else, we should have to go on accepting it. Even this should not blind us to its disadvantages. When thirty years ago I first began to look round for an alternative method of collecting manurial material, the simple reason was not the disadvantages mentioned, but that *there did not appear to me to be enough manure available* to the Indian peasant on whose behalf I was working. The national habit of burning the cow-dung as fuel severely limited what could be put on the fields, and I became convinced that some method of eking out his

206

scanty supplies was essential if he was to take advantage of the advances in plant breeding which the agricultural research workers of India were making: otherwise our work would be stultified. It was natural to study the successful methods in use in another part of the East and to consider the ideas underlying the Chinese practice of increasing the volume of fertilizing material by composting animal and vegetable wastes together. It quickly became part of my own routine to compost all the wastes of my experimental areas. The practical results soon forced themselves on my attention, but only in the course of time did the full meaning of the Chinese principles become clear to me.

In the end the substitution of the compost heap for the manure heap in my work proved to have been the most significant step in my education as a scientific investigator.

SHEET-COMPOSTING AND NITROGEN FIXATION

Subsoiling and the reform of the manure heap are the first steps in the solution of the problem of manuring. These will enable the soil to make a further supply of humus by a third method—sheet-composting. The fourth and last step naturally follows—the encouragement of the non-symbiotic soil organisms like *Azotobacter,* which fix atmospheric nitrogen.

Once the surface soil has been improved by the circulation of minerals and the supply of humus, the land will be in a condition *to begin to manure itself* by the process of sheet-composting. By this is meant the automatic manufacture of humus in the upper layers of the soil. Naturally the raw materials for this must first be provided. These are: (1) vegetable residues in the shape of the stubble and roots of crops like cereals; (2) temporary leys due for ploughing up, *which must always include deep-rooting plants and herbs;* and (3) green-manures, catch crops, and weeds.

For humus of the first quality to be made quickly from these three classes of vegetable matter we must always provide a supply of animal residues, either in the form of the droppings of animals or of reformed farmyard manure (compost). Besides this activating material we need oxygen, moisture, and warmth. If the land is properly farmed, we do not require a base to neutralize acidity: the soil will arrange this matter for us. Oxygen, of course, comes from the atmosphere and costs nothing:

the moisture is provided by the soil, by rain, and by dew: the necessary warmth is available if we begin sheet-composting before the land begins to cool in the late summer and early autumn.

The best results will aways be obtained with sheet-composting when the stubbles, temporary leys, green-manures, catch crops, and weeds are only lightly covered with earth. A deep covering of soil must be avoided, as sheet-composting requires a copious supply of air. The fermenting layer only needs just sufficient soil to keep the mass moist. When stubbles have to be converted into humus, *the supply of moisture can be enhanced by composting and lightly burying as soon as possible after reaping and before the surface soil has time to dry out.* There is nothing to prevent this operation following the binder once the sheaves are set up in rows, leaving narrow untreated strips between the cultivated areas.

Provided the soil is in good heart, a second composting is possible by sowing a catch crop on the sheet-composted land. Such land will do two things at the same time—prepare compost, and grow a catch crop. These catch crops can either be eaten by stock or disced in before winter comes. *The object of all this is to make the fullest use of solar energy by always having the soil in the late summer or autumn under a crop of some kind or, failing a crop, under weeds.* Vegetable matter must always be made and then converted into humus for the following year.

Proceeding in this manner a useful supply of humus will be created and ready for nitrification for the next year's crop. Further, all nitrates formed in the soil during the late summer and early autumn, which otherwise would be lost by leaching or denitrification, are immobilized and carried forward safely to the next crop.

Everything now will be ready for the last item needed in the solution of the nitrogen problem—nitrogen fixation. The organisms which carry this out must be provided not only with organic matter—to supply energy and food—but also oxygen, moisture, and a sufficient supply of base such as calcium carbonate to prevent an acid condition of the soil developing. It is more than probable that the good results which often follow dressings of chalk or powdered limestone are due in large part to nitrogen fixation.

Such fixation also takes place in a properly made compost heap; it must be continued in the soil; this is, however, only possible in really well farmed land.

The view that we must make every use of natural means—such as subsoiling, the full utilization of animal and vegetable wastes, sheet-composting, and nitrogen fixation—before even thinking of spending money on chemicals needs no argument. It will, I think, be found that when we make the fullest use of all these methods and follow the teachings of Mother Earth, we shall find it difficult to escape the conclusion that Nature, after all, is the supreme farmer.

THE UTILIZATION OF TOWN WASTES

The zones of agricultural land round our towns and cities are largely used to produce the fresh vegetables, fruit, and milk needed by the population. These areas ought, therefore, to be maintained in the highest possible condition. For this large volumes of compost will be needed. How is this to be obtained in areas where the supply both of vegetable waste and of activators of animal origin are certain to be small? The answer is: By the conversion into humus of the wastes of the towns themselves supplemented by baled straw brought in from outside.

Although our towns are fed from the countryside, little or no return of urban wastes to the land takes place. *The towns are, therefore, parasitic on the country. This will have to stop.* The wastes of these areas must go back to the soil. This can easily be accomplished by large-scale humus manufacture on the part of the municipalities. Instead of allowing the dustbin refuse to be buried in controlled tips or burnt in incinerators, this material should be turned into compost by the help of the crude sewage from the mains.

Two methods of using crude sewage as an activator are possible. We can either use it direct or filter it and then convert the sludge into powder, at the same time rendering the filtrate innocuous by chlorination. Both this dried sludge and crude sewage are excellent substitutes for animal activators. A small amount of dried sludge—about 1 per cent of the dry weight of the vegetable matter used—is sufficient to activate vegetable wastes. This powder will provide the owners of urban gardens and allotments with an excellent substitute for the animal manure now so difficult to obtain. The use of crude sewage is also practicable: long shallow pits may be filled with several layers of baled straw and dustbin refuse, which can then readily be moistened and activated by the sewage without the least nuisance and converted into excellent compost in some three months.

To get all this under way in Great Britain successful examples must first be provided to overcome the well-known inertia of our municipalities. Some are already in existence. In South Africa a nation-wide organization for converting the wastes of their towns and cities is in operation, as will be seen from the account contributed by Mr. J. P. J. van Vuren, the Co-ordinating Officer for Municipal Compost Schemes, in Appendix C (p. 274). The preparation of dried sewage sludge is described in an article by Dr. Lionel J. Picton, O.B.E., in the *News-Letter on Compost*, No. 10, October 1944. On page 239 there is a description of a method of converting straw into compost by means of crude sewage only.

Fortified by successful examples elsewhere and stimulated by the already growing demand for properly made humus, *it is only a question of time before our municipalities take up the preparation and sale of high quality compost* and show how the town can make some return to the soil to which it owes its life.

SUMMARY

1. The manurial problem can best be solved by copying the methods of Nature.

2. The circulation of minerals between subsoil and soil must be restored by means of afforestation and the subsoiler followed by the use of deep-rooting plants in the temporary ley.

3. The nitrogen problem can be solved by: (*a*) the reform of the manure heap; (*b*) by the sheet-composting of stubbles, green-manures, catch crops, and weeds; (*c*) by assisting the fixation of atmospheric nitrogen.

4. An ample supply of compost in the neighbourhood of towns and cities can be provided by introducing municipal composting on the lines now in successful operation in South Africa.

13

THE INDORE PROCESS AND ITS RECEPTION BY THE FARMING AND GARDENING WORLDS

THE SYSTEM of composting which I adopted, known as the Indore Process, has already been fully set forth in 1931 and 1940 in two previous books: [1] the detailed description will, therefore, not be repeated here. For those who are not familiar with these accounts it may be briefly stated that the process amounts to the collection and admixture of vegetable and animal wastes off the area farmed into heaps or pits, kept at a degree of moisture resembling that of a squeezed-out sponge, turned, and emerging finally at the end of a period of three months as a rich, crumbling compost, containing a wealth of plant nutrients and organisms essential for growth.

Sufficient time has now elapsed since the publications referred to above to permit of a summary of the history and reception of the process. The review is of interest. Time has sorted out essentials. It has brought no fundamental modification of any kind, but has shown the way to some simplifications which make the process easier both for the large plantation and for the small cultivator; it has indicated where further research and experiment could very advantageously be directed, and it has, above all, provided an interesting example of the way in which *a new presentation of a very old and well-tried idea has been warmly accepted by the practical man and given a most unfortunate cold shouldering by the leaders of agricultural education and research.*

Compost is the old English word for decayed organic wastes prepared by the farmer or gardener. There are many ways of making compost and it is a fact that, even when very imperfectly prepared, a heap of decaying organic material will, in course of time, turn into compost of a

[1] *The Waste Products of Agriculture; Their Utilization as Humus* (Oxford University Press, 1931).
An Agricultural Testament (Oxford University Press, 1940).

sort. There must be in existence dozens of indigenous methods of reducing the waste materials of Nature to nourishment for the plant: almost any traveller from primitive countries could describe some example. These empiric methods vary a good deal, mostly by reason of the different types of material available for composting. Actually *the basis is always the same, namely, to allow or induce microbial action by means of air and of moisture. It must never be forgotten that living organisms and not human beings are the agents which make compost.* These organisms exist everywhere. They prepare the ideal humus on the floor of the forest and they equally govern what goes on in the compost heap from start to finish. *The art of preparing compost amounts only to providing such conditions as will allow these agents to work with the greatest intensity, efficiency, and rapidity.*

The compost prepared by the Indore Process is like any other first-class compost. The method involves no patents, no special materials have to be sent for, and *there is nothing secret about it.* It is as well to make these points clear at the outset, as of recent years, owing to the immense success which has attended my compost campaign, numerous innovations and copies have been placed on the market, mostly patented and frequently involving the purchase of inoculating cultures or plant extracts of secret manufacture, some even claiming to be based on esoteric knowledge of an advanced kind and so benefiting the health and happiness of the recipient. Some of these have been described as a mixture of muck and magic. The Indore Process makes no claim of this sort whatever. *It merely copies what goes on on the floor of every wood and forest.* It has not been patented and will not be patented, because it would not be in accordance with my principles to make monetary profits out of work paid for from governmental and trust funds. Such results should always be public property and at the disposal of all. The Indore Process is now used and known in England, Wales, Scotland, and Northern Ireland; Eire; the United States of America; Mexico; Canada; Australia; New Zealand; South Africa; Rhodesia; Nyasaland; Kenya; Tanganyika; West Africa; India; Ceylon; Malaya; Palestine; the West Indies; Costa Rica; Guatemala; Chile; and by some of our armed forces. This list is constituted exclusively of countries from which I have directly received correspondence or official information.

It is because the Indore Process accords with natural law that it is equally successful in whatever type of farming or gardening it is applied. This is bound to be so. Nature has not different laws for her tropic, semitropic, temperate, or other zones, nor different principles for this soil or that. Her adaptations vary, but her basis is one and universal. It is a substantial proof of the soundness of the Indore method that it has shown itself to be successful in so many different climates and for all types of farming and gardening, and that nothing essential has had to be altered or added in the carrying on of the process.

The secret of this success lies in the quality of the product. *We must always secure high quality in compost before we can hope for quality and resistance to disease in crops, livestock, and mankind.* There is all the difference in the world between Indore compost and organic matter. This distinction is constantly forgotten by the apologists and supporters of the artificial fertilizer industry in criticizing organic farming and gardening, due, I believe, to want of first-hand experience of the subject.

SOME PRACTICAL POINTS

The objection is still occasionally brought forward that there is not enough material to compost. As was previously pointed out,[1] the true answer to this is a more effective use of the land. The proper utilization of the nitrogen cycle in Nature will provide much additional vegetable matter. There is also very considerable scope in the composting of catch crops and in sheet-composting generally. Sheet-composting has the added advantage that it saves labour, because the stubble or turf to be sheet-composted is *not* collected: it is left *in situ*. A parallel advantage is secured in respect of animal wastes when methods of open-air dairying like the Hosier system are adopted: obviously again the animal disperses its own wastes which mingle naturally with the vegetable wastes. All such methods need to be carefully studied as part of the fertility cycle; there is here an ample field for the intensive study of the nitrogen cycle and its full utilization in composting, and above all for pioneer adventures.

In any case, it may be insisted on once again that there is often a curious inability to recognize *the abundance of existing wastes*. The would-be complainant simply does not observe the many wastes lying about, the verdure of odd grass-borders for instance, the clippings of hedges; sometimes does not even see the weeds which encumber his beds and crops. One potential source of waste in this country is crim-

[1] *An Agricultural Testament*, p. 42.

213

inally neglected—the rich mixed growth along the sides of every country road in England. Quite frequently heaps of this growth are already well on the way to compost and need only to be removed. Systematic clippings twice a year (June and September) of the grass and weeds growing alongside the roadside hedges, ditches, streams, and canals would produce millions of tons of compostable materials. To save local authorities the labour and cost of clearance—for purposes of keeping the roads free the normal practice is to heap it up at the sides, a process which in itself must cost the country thousands of pounds per annum— is there any conceivable reason why the inhabitants of the localities should not be free to remove it for their own purposes? The riches of the roadsides and waste places would thus be brought back and add their wealth to our gardens and fields. This is not yet done, because this *nation has not yet been taught to look for and seize upon all available supplies of organic waste. Such training, nevertheless, is a national duty.*

In towns *the abundant autumn fall of leaves* which the authorities so carefully remove so as not to impede pedestrian and vehicular traffic and often destroy should be promptly returned to the gardens bordering on the roads so cleared; not to do so is year by year to rob these gardens of irreplaceable organic matter.

The condition of the soil receiving the compost is a factor fundamentally affecting results. This is only another facet of the problem with which we have just been dealing—the state of the soil which is to produce the compostable material. Run-down land produces little waste material, but it eats up compost at an inordinate rate. The first dressings seem to be sucked in at once: they disappear miraculously in a very short time. *The soil is so hungry that it positively devours compost.* But as the applications are repeated, the response of the crop is evident by a marked improvement in vigour, growth, colour, stance, foliage, flowering and seeding capacity. The cumulative effect is truly astonishing. The results of compost are soon written on the crop. Again and again in this country correspondents report that the mere appearance of a composted garden invariably attracts the attention of passers-by and secures new converts to organic gardening.

How can the new convert to organic gardening begin to obtain results? One method is to concentrate on *building up the fertility of the nursery* where seedlings are grown. The principles which have been so successfully applied to human infancy by the medical authorities of this country are true for plants also—at all costs give the seedling a good start. As soon as possible save the seed for future sowing from compost-fed plants. Provided the soil is fertile, the seed contains a whole battery of reserves. The next step is to sow such seed in soil rich in humus. The transplanted seedlings are then sure to prosper. This is the secret by which the rice cultivation of the East has been maintained for centuries year after year on the same land: the seed is carefully selected: the seedlings are always raised in heavily manured nurseries, and in this way survive the transplanting process on what they have accumulated. Or another simple method is to fill seed drills with two inches of compost and cover the sown seed with another inch: spectacular results, particularly with salad crops, can be obtained in this way. Or, again, in flower cultivation, when compost supplies are at the moment limited, a little compost may be poured into the site for the young plant or just round the roots of a growing one. All these devices are simple means of putting the compost where the crop in being can best use it. The ideal, of course, is to have the whole soil in such a state that any plant or seed can be set to grow anywhere without the need of special feeding. This, however, will take time.

The finished compost can be fed to the crop at any moment. In the more refined gardening operations it is a distinct advantage to possess a manure which can be spread on the surface to a depth of anything from one to two inches without the slightest disturbance of roots or seedlings. This is much nearer to Nature's own mechanism of distribution than is our common process of digging in at intervals raw fertilizing material which must necessarily be allowed to rot between the growing of crops, for which purpose ample time has to be allowed. In all intensive gardening operations compost is a necessity. A rapid succession of crops is thereby induced far surpassing what is permitted by other systems of manuring. Crops overtake each other, a second and third being interpolated while the first is ripening: the soil easily bears the double or triple burden. Here the Chinese peasant has led the way. No other agriculture is known which gets so much off the ground and has maintained unimpaired the fertility of the soil for four thousand years. Chinese agriculture, based on composting, is indeed the adaptation of genius, a marvellous achievement of a marvellous people, and would be well worth studying for its own sake even if it did not offer us such immense practical benefits.

How do we know when an area of land is really fertile? By the reaction of the crops to a complete artificial manure. When composting has been carried on for a sufficient period, soil which is in perfect heart does not respond appreciably to artificial manures—just as a body which is in perfect health ceases to show any marked reaction to stimulating drugs. When the soil is almost worn out, we can write our name on it with artificials, but as it becomes fertile the response to chemicals become less and less until finally no appreciable result can be observed. *The negative reaction of a treated area to a complete artificial manure will show that a condition of real soil fertility has been reached.* Here we must admit a useful, but somewhat restricted, opening for artificials. Once the land is in good heart the maintenance of fertility needs only moderate dressings of compost.

THE NEW ZEALAND COMPOST BOX

The rapid spread of the Indore Process in temperate countries with a well distributed rainfall has drawn attention to the advantage of providing adequate shelter for the small compost heap. The large heap will always protect itself, because the ratio of the amount of surface to the total volume is low and the mere size of the heap prevents any fall in temperature by the cooling effects of wind and rain. But a small heap is all outsides, so to speak, and is easily cooled. The fermenting mass, therefore, needs some protection. A simple method of providing this comes from New Zealand, where a compost box is now in use which is finding favour among the urban gardeners and allotment holders of this country. The best results are obtained with a pair of these New Zealand boxes side by side, the purpose of the second box being for ripening the compost.

Two suitable boxes can be made as follows. Both are exactly the same size, so the following description applies to both.

Materials required. Six 3 ft. 3 in. lengths of 2 in. by 2 in. for uprights. Twenty-four 4 ft. lengths of 6 in. by 1 in. board for the four sides of the box. The unplaned timber should be oiled with old motor oil to preserve it, but tar or creosote should not be used.

The box (see diagram), which has no bottom, stands on the ground. First nail the side A to the uprights E and F. Next nail the back B to the uprights G and H. Next nail the side C to the uprights I and J. When nailing the boards on to the uprights leave a half-inch gap between all boards to provide ventilation. The three sides of the box are now com-

216

plete. The sides and end are bolted together by means of four bolts—each fitted with two washers and a nut which unscrews on the outside—which join the back B to the uprights F and I. The front D is made up of loose boards, 6 in. by 1 in., slipped behind the uprights E and J as the heap rises. To prevent the sides A and C from spreading outwards use a wooden bar, 2 in. by 1½ in., with two wooden blocks, 3 in. by 2 in. by 1½ in., as indicated in the ground plan below of the box and the elevation of the bar K.

If the box has to be moved to a new site, remove the loose boards and the four bolts and re-erect the box in a fresh place.

Making the heap. Having made the box, throw your *mixed* vegetable material (broken or cut up if necessary into lengths a few inches long) into it as it comes to hand, together with one-third the volume of manure, *mixing the wastes and manure as the box is filled.* The proportion by volume of mixed vegetable wastes to manure should be three or four to one. All garden or unused kitchen waste may be used including weeds, lawn mowings, crop residues, leaves, hedge clippings, and seaweed when available. Where animal manure or soiled animal bedding is not available, activators such as dried blood, hoof and horn meal, or fish manure should be used, but in these cases only a very thin film is needed for every six-inch layer of vegetable waste. The exact quantity of these activators is 1 per cent of the dry weight of the vegetable wastes. If none of these substitutes for farmyard manure can be obtained, the heap can be kept moist—not wet and sodden—by means of bedroom slops.[1] Animal wastes in some form are essential. When urine earth is not used, sprinkle every six inches of the mixed vegetable and animal matter with a layer, about one-eighth of an inch thick, of earth (mixed with wood ashes, powdered limestone or chalk or slaked lime if available). A thin film to neutralize excessive acidity is all that is needed; too much earth hinders the ventilation of the mass. Then lightly fork over the layers of vegetable and animal wastes so that they get well mixed. This will help the fermentation and save the labour of turning.

If the wastes are very dry they must be watered with a rose tin till a condition like that of a pressed-out sponge is reached. If, however, about half of the vegetable wastes consist of fairly fresh green material, no extra watering will be needed. If a larger proportion still be green succulent stuff, it should be withered first and then wetted before use,

[1] If the bedroom slops are emptied each morning into a heap of good soil, all smell ceases in a moment and day by day the heap comes more and more to deserve the name of "urine earth" and is to be used in the box.

otherwise silage and not compost will result. A little experience will soon show how the moisture factor in composting should be managed.

Continue the building process until the total height is reached. After the box is half full make and maintain a vertical ventilation hole by thrusting a light crowbar or stout garden stake into the heap and working it from side to side. The hole should go as far as the earth underneath the box. The purpose of this ventilation vent is to improve the air supply.

The box should be protected from rain and sun by means of two pieces of old corrugated sheeting, each 58 in. by 26 in. These are kept in position by means of bricks or stones.

Two things must be watched: (1) an unpleasant smell or flies attempting to breed in the heap. This ought not to happen and is generally caused by over-watering or want of attention to the details of making the heap. If it occurs, the box should be emptied and refilled at once. (2) Fermentation may slow down for want of moisture, when the heap should be watered. Experience will teach how much water should be added when making the heap.

Ripening the compost. Provided due care is taken in filling the box, after six weeks or so the contents will be ready to be moved into the second box alongside (care being taken to place any undecomposed portions in the centre), the material should be watered if needed to keep damp, and allowed to ripen for a month or six weeks. No ventilation vent is needed for the ripening process. The compost which weighs about a ton is then ready for use and should be applied to the garden as soon as possible. If it must be stored, it should be kept in an open shed and turned from time to time.

In some places it may not be possible to find the wood or other materials—sheet iron or bricks—needed for the two bins. In this case two heaps side by side will serve, the method of assembly and turning being exactly as that described above where bins are available.

How much compost can be made in a year in a pair of these compost bins? At least four tons. We need never weigh compost. It can more easily be measured. As a general rule 2 cubic yards (54 cubic feet) of compost weigh 1 ton.

This simple device has been outstandingly successful. The speed with which material crumbles when protected by the New Zealand box from the outside cold is remarkable: a bare six weeks in the first box will often complete the active fermentation, after which the mass can be

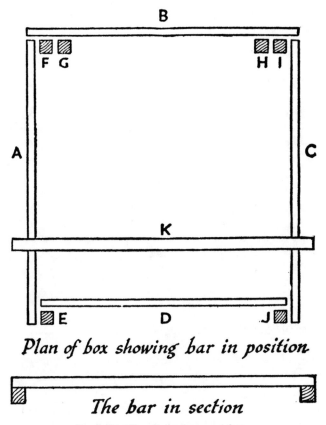

Plan of box showing bar in position

The bar in section

Fig. 5. The New Zealand compost box

A, B, and C are the sides, each consisting of six boards, each 4 ft. by 6 in. by 1 in., nailed to the uprights half an inch apart to allow ventilation.

D is the loose front (six boards).

E, F, G, H, I, and J are the uprights (each 3 ft. 3 in. long).

K is the bar, provided with a block at each end, to sit on top of the sides A and C to stop them spreading.

transferred to the second box for another six weeks for ripening. For those who have only small quantities of waste a pair of these boxes is just the adaptation required: they are extremely neat and tidy and take very little space. Proceeding in this way there is never any waste material left lying about. Household wastes can immediately be got rid of, and the composter may rest assured that neither flies nor smell will develop.

Local authorities might consider whether they could not provide such compost bins made of open brickwork as permanent garden fixtures in any post-war scheme for improved housing. The cost would be small and the advantage immediate and considerable, not least by definitely reducing the bulk and weight of the dustbin refuse to be collected: it is probable that the economy thus effected would soon repay the cost of this simple installation. The immediate and cleanly disposal of household rubbish is likely to make a strong appeal to every housewife and is a point worth study. Local authorities are spending large sums on the construction and upkeep of new houses. Why should not the maintenance of the fertility of the gardens round these houses be made a plank in our future housing schemes?

MECHANIZATION

The labour involved in making a small amount of compost is quite moderate: it is part of the routine of allotment holders and gardeners to keep their places tidy, and it is their usual habit to wheel their weeds and wastes to some special spot. To assemble this waste properly, add a little animal activator and soil, and when necessary do an occasional turn to the whole takes anything from a matter of a few minutes to an odd half-hour. It is fortunate that compost, by its nature, is not heavy, not nearly so heavy as ordinary manure; it can easily be handled by a woman. My wife turned a heap of about four tons in the course of two days without undue exertion.

But the work which the ordinary householder can take in his stride has to be differently considered by the farmer and the grower who pay for each hour of work expended on the farm or market garden. On this head many inquiries and some objections have been brought to my notice in the course of the last ten years. The original investigations made by myself and Mr. Wad were designed to assist the Indian cultivator. We did not concern ourselves very much about the factor of labour, for labour in countries like India is superabundant. In *The Waste Products of Agriculture* we stated (p. 13):

"Labour . . . in India is so abundant that if the time wasted by the cultivators and their cattle for a single year could be calculated as money at the local rates of labour a perfectly colossal figure would be obtained. One of the problems underlying the development of agriculture in India is the discovery of the best means of utilizing this constant drain, in the shape of wasted hours, for increasing crop production."

In Western agriculture, however, there is no such surplus of labour. In so far as I originally contemplated the use of the Indore Process in Western agriculture, I always looked forward to some form of mechanization as the best way of solving this problem. The recent advances which have been made in this direction and which will be described immediately below should not, however, cloak the fact that half the labour battle can be won by good management. It has frequently been noted by my numerous correspondents that the work involved in compost making can very largely be done not by the engagement of additional workers, but by a judicious disposition of the time of those already on the payroll. In any large-scale farming enterprise there are off hours which can be advantageously used for compost manufacture. For instance, the collection of material, which is a big item on a large estate, can be made a matter of arrangement of carts and men on their return journeys. Obviously the site for the compost heaps or pits needs to be carefully determined with a view to the shortest journeys both for bringing in the raw material and for carrying out the finished product. At the Indore Experimental Station the composting pits were placed next to the cattleshed in the centre of the whole area. In any case, as some of my correspondents early pointed out, the labour expenditure may prove well worth while for an operation that so notably adds to the capital value of the estate, as well as contributing to the profit and loss account.

Giving due value to all these considerations, nevertheless the question of labour remains of obvious importance. In two directions the situation has turned out very promising. In the first place, experience has proved that my original estimate of the need for turning the compost heap three times was excessive: one turn, or in very disadvantageous conditions (e.g. excessive rainfall) two, is all that is necessary. The experience of my correspondents, and my own further personal experience in making small compost heaps, places this fact now beyond doubt and it is a very great gain in economizing both time and labour.

The secret of correct compost making has proved to be mixing the ingredients at the outset and attention to the aeration of the fermenting mass. Provided this is done, a single turn is sufficient. Even without a turn well mixed and well aerated material will decay fairly well. The methods used in aerating large heaps since the original experiments at Indore have been described in *An Agricultural Testament* (p. 235 *et seq*.). Mr. Dymond in Natal has devised another simple method of supplying air from below the fermenting mass which is certain to be widely adopted (p. 226).

Better mixing and improved aeration thus eliminate repeated turnings. Assuming, however, one turn is necessary, how is this to be done with a minimum of labour? There is also the question of loading and spreading the finished compost. The problem applies particularly to large-scale work in Great Britain. As already indicated, the solution is bound to be by means of some machine so devised as to be capable of performing the three operations of assembling, aerating, and loading. A great deal of progress has been made in this direction. Mr. Friend Sykes of Chute Farms Limited, Chute, near Andover, has invented a muck-shifting crane driven by a caterpillar tractor. This is described in Appendix D. He has also invented a simple manure distributor. A number of other machines for compost making have been devised, so an interesting contest between the rival machines will soon be taking place. That machine which will stand up to the work and also produce high quality compost will win the battle. That so much attention is now being paid by inventors and manufacturers to the mechanization of compost making speaks volumes for the progress organic farming is making.

Mr. Sykes' muck-shifting crane, which has been made by Messrs. Ransomes & Rapier Limited of Ipswich, will turn and aerate a compost heap and also load the finished compost into a manure distributor. I understand that this machine will load 200 tons of muck in a day at a cost, including spreading, of 1s. 8d. a ton. These operations cannot be done by hand labour under 12s. 6d. a ton. If such savings can be realized in general farming practice, organic farming by means of the reformed muck heap is certain to prove much more economical than present-day farming with the help of the fertilizer bag.

The proof of the pudding is always in the eating thereof. An interesting and even exciting contest between the disciples of Rothamsted and the humus school is certain to develop. In such a struggle the verdict must inevitably be given by the crop and by the livestock and not by the lawyers on either side.

THE SPREAD OF THE INDORE PROCESS IN THE FARMING AND PLANTATION WORLDS

The Indore Process was first taken up by a number of pioneers in the farming and plantation worlds like Colonel Grogan in East Africa, Captain Moubray in Rhodesia, Colonel Sir Edward Hearle Cole in the Punjab, the late Sir Bernard Greenwell and Mr. James Insch in England, who, undeterred by the criticisms of the experts, started out to test the process and then to initiate a large-scale composting programme on their properties: their success was immediate: the spread of the composting principle was inevitable the moment my ideas began to be written on the land. Their efforts have also attracted the attention of some of the public authorities in their respective countries who have been quick to avail themselves of those developments in the Indore Process which lead towards new advances in crop production, in sanitation, and in public health. In Costa Rica, Señor Montealegre, first in his capacity as Director of the Institute for encouraging coffee growing and second as Minister of Agriculture and Lands, has spared no pains in making my work known throughout Latin America. Another stage was soon reached when a number of allotment holders in England began to approach me for advice and help: the spread of composting in the smallholding, allotment, and private garden is not the least useful of the developments in the compost campaign. I have naturally done all in my power to encourage and help these pioneers and to discover still more pioneers. It is to the work of these men and women, especially to the early advocates of composting, that the spread of the humus idea is due.

Full details of the progress made up to 1940 will be found in Chapters V to VIII of *An Agricultural Testament*. In the short period which has elapsed since, a number of facts confirmatory of the principles which I have advanced have been brought to my notice from many countries: much of this information will be found in the twelve issues of the *News-Letter on Compost* [1] from October 1941 to June 1945.

[1] Published by the County Palatine of Chester Local Medical and Panel Committees at Holmes Chapel in Cheshire at an annual subscription of 5s.

The most recent advances in the application of the composting idea can best be described country by country rather than crop by crop: much of the new work has been on the general composting of wastes and urban refuse or has taken the form of the setting up of organizations to further the principles involved. The attention of the press has been awakened; the compost heap has even crept into the cartoon. The medical and educational professions are becoming increasingly interested, and there is every sign that an avalanche of converts is rapidly threatening to sweep away such opposition as is based on ignorance, apathy, or vested interests. In the succeeding sections of this chapter a few of the more outstanding developments of the last four years are summarized.

SOUTH AFRICA

Reference was made in *An Agricultural Testament* [1] to the assistance given to my theories by the work of Mr. Dymond, Chief Chemist to the South African Sugar Company in Natal. Mr. Dymond supplied me with abundant material in the form of roots of the sugar-cane, grown with artificials only, with humus only, and with both. From these samples Dr. Levisohn established the fact that the sugar-cane is a mycorrhiza former and that artificials were injurious by preventing the roots from digesting the invading mycelium: where humus was used, there was abundant mycorrhiza formation and rapid digestion of the fungus.

These results suggested that the change-over from pen manure (a rough form of farmyard manure) to artificials lies at the root of the diseases of the cane and is the cause of the running out of the variety. We seem to be dealing with the consequences of incipient malnutrition —a condition now becoming very general all over the world in many other crops besides sugar-cane. Interesting confirmation of this view has now been obtained by Mr. Dymond. In 1938 an experiment was commenced to study the effect of compost on streak disease (a virus trouble) in Uba cane. A few plants of moderately virus-infected cane were planted in a short row with a normal dressing of compost. During the following two years there was no increase in the disease which was estimated at 60 per cent. In the meantime the original plants developed a 100 per cent infection. After the second cutting the ratoons were surface dressed with fresh compost. At the end of the third year the disease had

[1] *An Agricultural Testament*, pp. 69–70.

diminished to approximately 25 per cent and during the fourth year the new growth was examined and passed as entirely free from streak.

Since then cuttings from the canes which have recovered from streak have been planted out in a composted seed bed, where they have so far maintained their immunity. A row of 100 per cent streak cane has been planted adjacent to this plot. No infection of the virus-free cane has so far developed after six months' contact.

Samples of the roots of the streak-diseased and streak-free (after four years' treatment with compost) canes were examined by Dr. Levisohn who reported no mycorrhizal infection in the former, but sporadic infection of the endotrophic type of fungus in the fibrous roots of the latter.

Dymond (*Proceedings of the South African Sugar Technologists' Association,* 1944) concludes his account of this valuable piece of work in the following words:

"The mycorrhizal association, after compost treatment of the virus-diseased cane, is significant and important, as it confirms the mycorrhizal theory and association in respect to sugar-cane.

"The streak-free Uba is growing vigorously and compares well with the deteriorated Uba fields common in the last ten years.

"The point to be emphasized as the result of this experiment is not so much that streak-free Uba cane may stage a come-back and provide a standby variety, but that the fundamental principle of soil fertility and the practice of the fertile seed bed may be applied to any suitable variety of sugar-cane. In this way only can the industry be assured of healthy seed and healthy crops in perpetuity."

It follows from the above that the direction in which the sugar industry of Natal can be placed on firm foundations is to manufacture as much compost as possible and to use this for growing the plant material.

Steps have been taken to devise a simple means of doing this. Following up the preliminary experiments on composting the wastes of the cane,[1] Dymond has just published a detailed account of a simple scheme for converting the night-soil of the labour force and the various sugar-plantation and factory wastes into humus (*Proceedings of the South African Sugar Technologists' Association,* 1944). The scheme is now in successful operation at Springfield Estate, Darnall, Natal. The results are so important and so far-reaching that a detailed account is essential.

At this estate a set of compost bins has been designed to promote the

[1] *An Agricultural Testament,* pp. 68-71.

easy filling of the pits and the removal of the fermented product for ripening. Each bin is provided with adequate drainage and abundant aeration. The capital cost of the lay-out is low, so that it can easily be adapted to the smallest farm or the largest factory or township. The plan and photographs (Plates V and VI) show the essential details of construction and the method of working.

The bins are built on sloping ground by means of hollow cement blocks and cement mortar. The concrete floor has sufficient slope for drainage and is provided with three longitudinal tiers of bricks to support a loose platform of bamboos or light poles, so arranged as to leave about an inch space between each pole for aeration. In this way the fermenting mass obtains abundance of air from below. The lower end of the bin is closed by a loose gate of poles held in place by two vertical pipes embedded in concrete.

For an annual output of 1,000 tons of finished compost, six of these bins, each 20 feet long and 9 feet wide by 4 feet 6 inches deep (810 cubic feet), are necessary. Such an installation will deal with the wastes of 250 people, 45 animals, 100 tons of filter press cake, together with the necessary amount of megasse, cane trash, and cane tops.

The method of operation is first to cover the poles with a light foundation of weathered cane trash and then with an eight-inch layer of cane trash or megasse which has been used for the bedding of livestock and which is impregnated with urine and dung. The next day the contents of the night-soil buckets are distributed over the absorbing mat. These are immediately covered with stable litter and the whole enclosed in a thin layer of filter press cake. The process is repeated every day. Light dustings of finely ground agricultural lime and applications of diluted molasses (50:50) improve the intense fermentation which sets in. Sufficient water must be applied while filling the bins to keep the material wet and to prevent drying out owing to the high temperatures reached which often touch 78° C.

The night-soil buckets are layered with megasse as an absorbing medium and covered with the same material on removal. Two long planks over the top of the bins facilitate charging and also avoid trampling and consolidation. The bins are filled about one foot above the surface as after a month the mass contracts to about two-thirds.

The pits should be filled in ten days and allowed to remain for six weeks. The partially rotted material is then turned out through the open end of the bin and allowed to ripen in heaps for another six to eight weeks, when it is ready to apply to the soil.

While the best method of using this installation to produce the most satisfactory compost has not yet been settled, the following analyses are interesting and tell their own story.

ANALYSIS OF COMPOST, SPRINGFIELD ESTATE, NATAL

	1	2	3	4	5	6	Karoo manure sample
Moisture per cent	69.8	61.3	69.0	63.8	77.0	78.0	36.8
Loss on ignition	45.8	29.7	38.1	34.8	59.6	59.6	47.9
Nitrogen, N.	1.7	1.0	1.2	1.3	2.2	2.2	1.7
Phosphoric oxide, P_2O_5 total	2.0	1.6	1.4	1.3	2.2	1.5	1.5
Phosphoric oxide, P_2O_5 available	0.7	0.6	0.6	0.8	1.7	1.2	0.6
Potash, K_2O total	3.8	1.2	2.7	1.0	1.1	1.7	10.7
Potash, K_2O available	1.3	0.5	0.9	0.7	0.6	1.4	3.8

1. Represents stable litter with cane tops, filter press cake, megasse, and old manure.
2. Represents the same with the cleaning-up of the premises.
3 and 4. Normal practice as described above, together with diluted molasses.
5 and 6. Normal practice with dustings of agricultural lime: no molasses.

The high percentage of nitrogen in 5 and 6 suggests that dustings of agricultural lime may favour nitrogen fixation. When the best method of procedure at Springfield has been devised, a nitrogen balance-sheet of the whole heap would make interesting reading. If matters can be so arranged that nitrogen fixation does take place, a new chapter in the manuring of the sugar-cane will have been opened.

As regards the sanitary aspects of this method of activating the wastes of the cane with animal manure and night-soil, the local Medical Officer of Health reported that he found no flies, no smell, and no nuisance. Pathogens could not possibly survive the conditions of high temperature and high humidity which obtain for many days in these bins. The method, therefore, combines two things: (1) the systematic removal and sanitary disposal of all the wastes of a sugar estate, and (2) the production of a valuable organic manure at a low cost.

In concluding his paper Dymond deals with future possibilities and the best method of utilizing the surplus vegetable wastes of sugar estates for the manufacture of compost in towns and cities. The average sugar estate produces an abundance of vegetable wastes over and above those that can be activated by the animal and human wastes now available. Thus from an annual crop of 6,000,000 tons of cane the following quantities of vegetable wastes are produced:

	tons
Cane trash	1,200,000
Cane tops	540,000
Megasse	1,980,000
Filter press cake	270,000
Molasses	180,000
Total	4,170,000

If these wastes were baled and transported to the towns and cities, a portion of the large quantity of vegetable matter needed for municipal composting would be provided.

As regards the sugar industry this Springfield experiment solves the humus problem. It will provide the large quantities of compost needed for producing the plant material for the succeeding cane crops. As the livestock population on these estates increases more and more humus will become available for the current crop.

It is a particularly happy circumstance that this great advance should have been made by a chemist. It makes the fullest reparation for the harm done by some of the chemists of the past through slavish devotion to chemical analyses and will also go a long way in emancipating future investigators for sugar-cane problems from the thraldom imposed by the NPK mentality. By regarding the manuring of the cane as a biological, as well as a chemical, problem Dymond has achieved a notable advance and one that is certain to be taken up far and wide. It is another milestone on the road to organic farming.

Just as this book was going to press, Dr. Martin Leake drew my attention to a note in the *South African Sugar Journal* of September 1944 on composting practice on the Tongaat Sugar Company's estates in Natal where noteworthy progress has already been made in converting the wastes of a sugar estate into compost.

This group of estates cultivates 16,000 acres of cane and manufactures 70,000 tons of sugar annually with a useful by-product in the shape of 18,000 tons of filter press cake.

The problem of maintaining the organic matter content of the soil is being solved by composting the cane trash and filter press cake together in heaps eighteen feet wide and five feet high. The aeration of the fermenting mass takes place naturally, as the mixture is sufficiently porous: moisture is supplied by rain. Two turnings are given and the finished material is used at the rate of thirty tons to the acre in the furrows for the new plantings on light land, the cuttings being laid on top of the compost. No animal activator appears to be used in these heaps, an omission which is sure to be rectified when more livestock is kept on these estates.

Green-manuring with *san* hemp is the rule on all the newly planted areas so that by this means and the compost placed in the furrows the supply of organic matter should be sufficient.

The animal residues of the estate oxen, horses, and mules are used to activate large quantities of cane trash in pens, the soiled bedding being afterwards converted into humus in the ordinary way, the yield working out at twelve tons per head of stock. This material is used mostly on the heavy lands.

In these two ways from 40,000 to 50,000 tons of compost are made annually by this enterprising company.

Last season the average yield of cane per acre on these estates was 45.88 tons, which is 60 per cent more than that of Natal as a whole. It is expected that when the full effect of the composting programme outlined above is obtained, considerably greater yields will be reached during the next few years.

The cane-sugar industry all over the world will naturally follow the pioneering work in progress in Natal both on the Springfield and the Tongaat Estates. This work on the conversion of the wastes of the cane into humus, coupled with the results the late Mr. George Clarke obtained on green-manuring and trench cultivation at Shahjahanpur in the United Provinces, is certain to place the cultivation of the cane in a truly impregnable position for many years to come.

The story of the composting of human wastes is continued in the notable pioneering work of Mr. J. P. J. van Vuren, which began at Ficksburg in the Orange Free State with two compost pits in 1939. Mr. van Vuren at once showed how the various wastes of a small township could be converted into humus by the Indore Process and the product sold to the farmers and gardeners near the town.

The population of Ficksburg is 2,750 Europeans and some 3,000 Natives. Soon eight compost pits were in operation, which at first produced about twenty tons of compost a month from such wastes as straw, leaves, waste paper, old bags, sawdust, shavings, wood-wool, weeds, hedge and lawn cuttings, stable manure, kitchen waste, wood ashes, abattoir wastes, and night-soil. These town wastes are collected by the municipal dust and night-soil carts and taken to the compost pits, which are a little way out of town.

The pits, which are now four feet deep, have brick walls with a floor slightly sloping towards the centre, where there is an aeration channel covered with bricks laid open jointed, and carried up at the ends into chimneys open to the wind. By this means air permeates the fermenting mass from below.

In filling the pits care is taken not to lose any liquid by providing a thick layer of absorptive refuse in the bottom of the pit, when the first load of night-soil is turned in and evenly spread; the method of charging carefully follows those set out in Appendix C to *An Agricultural Testament*. The fermenting mass is turned twice, the entire process taking from eight to ten weeks, depending on the type of material used. There is no odour from a pit properly filled, because the copious aeration effectively suppresses all nuisance.

In Ficksburg the compost is sold to farmers of the district for use on their lands or orchards and in town to local gardeners and private individuals for use on their lawns and gardens. The farmers send their waggons and take delivery at the compost pits, but in the case of smaller orders these are delivered by cart, either loose or in bags. Repeat orders are numerous because the crops in the district, as well as many gardens and lawns, have proved excellent advertisements.

The result of this one successful example of municipal composting was immediate. One practical example worked wonders. Other municipalities—Volksrust, Heidelburg, Bethlehem, Hercules, Walmer, and others —copied it; still more became interested. Soon a scheme covering the whole of the Union of South Africa was under way. The Union Government appointed Mr. van Vuren as Co-ordinating Officer for Municipal Composting and divided the area under their jurisdiction into six regions, each in charge of a composting officer. Progress has been rapid and now the urban wastes of many of the large towns are being converted into humus for the benefit of the neighbouring farmers and gardeners. A detailed account of the progress of this nation-wide municipal composting scheme will be found in Appendix C to this book. From the municipalities the work of humus production has spread to the countryside and Mr. van Vuren now has a colleague for dealing with humus production on the farms.

It is to Mr. van Vuren also that I owe confirmation of my statement about the possibilities of improved wine production from fertile soil, the only road of escape from the threatened dangers of disease, loss of quality, and the running out of the variety.[1] In a letter dated 5th May

[1] *An Agricultural Testament*, pp. 85-6.

230

1944 he informs me that he has found an example of wine production from fertile soil near Capetown. At the Nederburg Farm, Northern Paarl, Western Province, Mr. J. G. Graue raises his grapes with organic matter only without any help from artificials. His wine, known locally as Nederburg Riesling, enjoys a high reputation for quality in South Africa. More such examples are urgently needed both from South Africa and Australia before our Empire-grown wines can come into their own.

It is not too much to say that the whole of South Africa has become compost-minded. All the preliminary work needed in blazing the trail has been done and local examples abound showing how the soils of this vast area can be restored to fertility. A great impetus has been given to this work by the recent formation of the National Veld Trust, who have made humus an important platform in their programme. The following article, which appeared in the issue of the South African *Farmer's Weekly* of 19th April 1944 (p. 235), explains itself:

COMPOST CLAIMS OFFICIALLY ENDORSED

A Fundamental Necessity for the Maintenance of Production

"In the course of its report to the Government the Reconstruction Committee of the Department of Agriculture says that in addition to sound methods of rotation, it is equally essential that all available plant and animal wastes be constantly returned to the soil in order to replenish its humus supplies and at the same time restore to it a substantial proportion of the plant nutrients taken up by the crops harvested.

"This is a fundamental necessity for the permanent maintenance of a high level of production and is all the more necessary in building up the fertility of old, depleted lands. It is the logical and natural method of fertility maintenance that has been followed through the ages in older countries, although it has suffered considerable neglect during the last few decades since commercial fertilizers have come into wide use.

"Happily there is a growing realization all over the world to-day that the use of fertilizers in no way compensates for lack of soil humus and that the full utilization of farm wastes as sources of humus must form an integral feature of the system of land use as a whole, a fact that applies equally to dry land as well as to land under irrigation.

231

Most Effective Method

"All experience goes to show that by far the most effective method of returning farm wastes to the soil is in the form of well-prepared compost. Alternative methods are by the direct ploughing in of untreated crop residues, by green-manuring, and the accumulation of animal manure in kraals or manure heaps for ultimate return to the land; but certain disadvantages attach to each of these alternatives as compared with the use of compost.

"Under farm conditions the limit to the amount of compost that can be made is often set by the supply of plant wastes available. Crop residues alone will hardly furnish enough material and, as a general rule, main reliance has to be placed on old veldt grass, mown for this special purpose.

"Where the supply of veldt grass is also strictly limited, the only remaining alternative is to grow bulk-producing grasses (on such spare area as may be available and also along fences and on contours between lands) as a source of compost material.

Cost

"On the basis of a meticulous costing of every operation involved doubt is sometimes expressed as to whether compost making pays. It is overlooked that the making of compost can hardly be regarded as an optional matter in cropping areas and that the normal farm routine can frequently be adjusted to include this activity with the employment of little additional labour.

"In practice, the actual cost of compost to the farmer is not only small, but should be amply recovered in the form of improved soil fertility.

"The time is rapidly drawing near when fruit and vegetable growers, who rely largely on supplies of kraal-manure imported from other parts of the country, will have to become self-sufficient in this respect and to produce their own requirements in the form of compost. This is the ideal at which every farmer should aim, where crop production plays any significant role."

One further fact from South Africa is of interest. To the account on maize published in 1940 [1] can now be added the evidence that maize, like sugar-cane, is, as was expected, a mycorrhiza former and is therefore

[1] *An Agricultural Testament,* p. 78 *et seq.* and p. 166.

232

provided with the means by which protein can circulate between soil and crop. Regular supplies of freshly prepared humus are, therefore, vital for this crop. Besides maintaining the crumb structure and the life of the soil, it assists the maize plant to resist all kinds of pests.

RHODESIA

Starting from the farms of the pioneers, composting soon spread in Rhodesia and now the Agricultural Department publishes every year a return of the number of cubic yards of compost made on the farms. In 1940 there were 674 farmers making compost; in 1943 the number had increased to 1,217. In the same years the amounts of compost made were 148,959 and 328,591 cubic yards. It will be seen that compost-making is going up by leaps and bounds, but the figures do not tell the whole story, as numberless small composting centres and private gardens are not included in the return.

The position is well summed up in the following extract from a letter from Captain Moubray to the Editor of the South African *Farmer's Weekly* (26th April 1944, p. 270):

"If we had realized the all-important role of humus years ago, and had acted on that knowledge, much of to-day's damage could have been averted.

"Even to-day there are those who are not satisfied that there is sufficient scientific proof that the basic principle involved in Sir Albert Howard's Indore Process of converting animal and vegetable wastes into compost or humus is a cure for many of our soil ills. Farmers in increasing numbers are, however, finding out for themselves, and when they see the results of compost on their lands they are not inclined to pay much attention to anything else.

"When Sir Daniel Hall visited Mashonaland some years ago, he quite refused to take Sir Albert Howard's claims seriously; but the small snowball of those days has, at least in these parts, become an avalanche sweeping everything before it."

This quotation, together with that given on p. 230 above, leave no doubt about the general results of the humus campaign, which began in 1932 when the *Farmer's Weekly* reviewed at length *The Waste Products of Agriculture* and afterwards opened its columns to a discussion, often very lively, between the local representatives of the artificial manure industry and the champions of organic farming. One result of this pub-

233

licity was to stimulate the pioneers to convert the waste products of their farms into compost and to observe the results. From that moment artificials began to lose the battle. Then the advocates of artificials changed their ground and took up the position that the soils of South Africa would best secure the restitution of their manurial rights by humus supplemented by sufficient artificials to produce a balanced manure. In this way they hope to stem the onward march of humus and to postpone the evil day when both the farmers and the urban dwellers in South Africa, as well as the purchasers of their exported agricultural produce, realize that the slow poisoning of the life of the soil is one of the greatest calamities that has befallen agriculture and mankind.

The onward march of progress in the Rhodesias owes much to Captain Moubray who for many years has written the results of humus on his farm and so provided the country with a successful example. I have done everything in my power to persuade the artificial manure interests how valuable it would be in their advertisement campaign to take up a piece of land next to Captain Moubray's estate and to show that by means of artificials, or artificials and humus, they could do even better. But they have preferred to lose face by declining the challenge rather than to risk a disastrous defeat. Discretion has proved to be the better part of artificial fertilizers.

In the early days of 1933 I paid a brief visit to Natal and South Africa and saw for myself how dire was the need for more humus. Just over twelve years have passed, but what a change has taken place in that brief period! I could, in 1933, discover but faint interest in humus and soil fertility among the people I met. To-day the virtues of humus are being preached everywhere: the purpose of the Indore Process is being widely understood: the flow of ridicule and abuse from the artificial fertilizer industry is coming to an end. I have enjoyed this battle with the protagonists of the NPK mentality: I have enjoyed still more a long and detailed correspondence with the pioneers, without whose labours nothing could have been accomplished in Rhodesia and in South Africa.

MALAYA

For some years before the fall of Singapore Malaya was one of the most active composting centres in the Empire, thanks to the enthusiasm of Dr. J. W. Scharff, the Chief Health Officer at Singapore, and of a number of men engaged in the plantation industries.

234

Composting began in Malaya on a number of coconut and rubber estates. An example of the kind of results obtained is given in the following letter dated 17th October 1941 from Mr. R. Paton, Permatang Estate, Banting, Selangor:

"We started to keep livestock on a fairly big scale in 1930 for the purpose of manuring our coconuts, and this was done in conjunction with composting of husks, fronds, etc., in trenches two feet deep along the centre of each row. These trenches were originally cut as surface drains, and as such they still function, while at the same time absorbing rainfall and providing moisture for the palms during periods of dry weather. Our average yield per acre was below nine piculs of copra, and the palms were then beyond the age at which one would expect any appreciable response in yield. Nevertheless, they have yielded over fourteen piculs per acre average for each of the past five years, and look like doing even better. Fine results have been obtained also in our rubber trees, particularly in young replantings, where the growth is all that could be desired, and not one ounce of artificial fertilizer has been used."

The great principle that the plantation industries can never succeed without livestock and properly made compost is well illustrated by the above experience. I observed the same thing in 1938 with coconuts in the low country of Ceylon—far healthier trees and much better yields where animals were kept in the groves. The outstanding weakness of the rubber estates I visited in South India and Ceylon was the total absence of livestock among the mature trees and no provision for making compost for the nurseries. It was little wonder that so much disease occurred.

But the most spectacular advances in composting in Malaya are due to the interest and enthusiasm of a number of medical men who were quick to grasp the possibilities of composting. Dr. Reid of Sungkai did much to make the ideas in *An Agricultural Testament* known to the planting community. Dr. Scharff, who first came in contact with humus at a lecture I gave in 1937 at the London School of Tropical Medicine, immediately after returning to Malaya took up the process, systematized it, and established it at Trengganu, and by means of his staff and his medical colleagues got it under way in Penang, Kelantan, Sarawak, and the State of Johore. Municipal composting was well established in Malaya before the Japanese invaded the country.

Full details of Dr. Scharff's composting campaign in Malaya were published, as the work developed, in the *News-Letter on Compost,*

235

No. 2, February 1942, pp. 2–9, and No. 4, October 1942, pp. 46–9. At an early stage it was found necessary to systematize composting and this took the form of the Trengganu Household Composting plan. The work was done within a fenced enclosure made of bamboo or jungle saplings four feet high. Four compartments were arranged for at one end of the enclosure and each of these compartments was filled with material during four successive weeks. Turning was done almost automatically and with the correct time spacing. Plate VII illustrates the lay-out and shows the position of affairs at the end of each month. The Trengganu plan was soon adopted all over Malaya. This was the position when Malaya was invaded by the Japanese. But before Singapore fell Dr. Scharff managed to complete a large-scale trial of compost-grown food on the Tamil labour force employed by the Health Department, already described in full in Chapter X of this book (p. 178).

INDIA

A very promising development in compost making is now taking place in India. Although an account of the Indore Process was published in 1931, nevertheless twelve years have elapsed before any official notice was taken of the possibilities of the compost idea. The direction this is now taking will be clear from the following letter addressed to me and dated 24th August 1943 from Dr. C. N. Acharaya, Chief Biochemist, Imperial Council of Agricultural Research, India:

"You will be interested to know that the Government of India have recently launched an all-India scheme for the preparation of compost-manure from urban refuse and have sanctioned an allotment of about 2½ lakhs of rupees for the purpose. The scheme is to be operated by the Imperial Council of Agricultural Research, and the above grant would be apportioned among the different Provinces and States in India for the purpose of training special officers (Provincial or State Compost Biochemists) in the technique of compost-making from urban wastes, and for organizing the preparation of compost-manure at selected municipal centres in the respective Provinces and States. I have the honour of being selected for the office of Chief Biochemist to the Imperial Council of Agricultural Research, who would be in charge of training the Provincial and State biochemists and, later, in supervising their work. The headquarters of the new scheme have been established at Nagpur, being geographically a central place, from which easy access could be had to all parts of India.

"As I am getting together all available literature relating to compost and organic manures for passing on the information to the Provincial and State Biochemists working under me in all parts of India, I should very much value it if you would kindly let me have available copies of all your papers and lectures on the subject, in addition to the publications issued by the County Palatine of Chester Local Medical and Panel Committees."

Although the Indore Process was primarily devised for the benefit of the cotton growers of India, whose interests are being looked after by the Indian Central Cotton Committee, little can be added to the section on cotton in *An Agricultural Testament* which carried on the story to the middle of 1940. No change appears to have been made in the research programme of this body. The obsolete idea that the problems underlying cotton production in India can be solved by plant breeding and the control of pests still holds the field.

One promising piece of pioneering work on cotton in the Punjab has, however, continued to develop on Colonel Sir Edward Hearle Cole's estate at Coleyana in the Montgomery District. Sir Edward is more than ever convinced of the value of freshly prepared humus for this crop. He finds that compost not only increases the yield, but improves the quality of the fibre as well. More large-scale examples like this are needed to confirm the view that the restoration and maintenance of the fertility of the soils producing cotton lie at the foundation of all progress in this crop.

NEW ZEALAND

In this Dominion the creation of pastures by deforestation followed by the excessive use of chemical fertilizers, superphosphates in particular, soon led to the rapid exhaustion of the land. Soil erosion is increasing; vegetables have lost their taste; the health of livestock is deteriorating. The more far-seeing of the population have been alarmed by the growing signs of malnutrition and the increase in the number of patients in hospitals and asylums, hence the formation of the New Zealand Humic Compost Club, the object of which is to encourage the fertilization of the soil by means of humus made from vegetable and animal wastes and so foster plant, animal, and human health.

The progress of this novel undertaking has from its inception been remarkable. Starting from small beginnings in 1941, by 31st March 1942 the membership was 440; a year later it was 2,007, and on 31st March of the present year (1944) it had reached 4,396, truly an amazing achievement, and one which reflects, on the one hand, the tremendous local interest in the vital principles of soil fertility advocated, and on the other, the successful manner in which the President, Dr. Chapman, and the Honorary Secretary, Mr. T. W. M. Ashby, have guided the new movement. The Compost Club has recently been incorporated as a non-profit-making company. It publishes a magazine—*Compost*—every two months and during the year ending March 31st last no less than 48,878 copies were printed and distributed. Besides the magazine a number of pamphlets have been issued, two of which have already passed the 20,000 mark. The Club also maintains a reference and lending library, and acts as a distributing agency for books printed overseas. There are ten local branches which arrange meetings, demonstrations, and field days. The Club finances itself from a small annual subscription of 5s. and is beginning to build up a substantial credit balance. Full details of this interesting development can be obtained from the Hon. Secretary, New Zealand Humic Compost Club Inc., P.O. Box 1303, Auckland, New Zealand.

The activities of this Club have not escaped the usual opposition, criticism, and even abuse on the part of the chemical fertilizer interests and their supporters, but this young organization is well served by a very able executive who has deftly used these attacks to advertise the new movement and to make clear to the population of New Zealand the immediate and the future issues involved in the restoration of soil fertility. When the time comes for the prodigal to return and to confess, the Compost Club will have ready to hand example after example showing the road out of the abyss into which New Zealand has fallen, by the simple expedient of the restitution of the manurial rights of the soils of the country. To-day the members of this Club are being described as a set of cranks: to-morrow they will be recognized as the saviours of their world.

GREAT BRITAIN

Even in this expert-ridden island of Great Britain and in spite of the additional restrictions imposed by the Defence Regulations new ground is constantly being broken by the pioneers—in farming, in gardening, and in nutrition.

In farming the chief advances have been made in two directions—in preparing the soil for additional humus by means of the subsoiler, and in the mechanization of the manure heap. These two important steps have already been fully described. A still more recent advance—an improved muck spreader—is referred to in the *News-Letter on Compost,* No. 10, October 1944.

These various labour-saving devices are leading to still further advances by which two important residues, now largely running to waste, can be used in compost manufacture. The first of these residues is straw, vast volumes of which now litter the countryside. These cannot be trodden down and converted into humus under the feet of live stock, because the supply of animals has not kept pace with the areas devoted to cereals. War farming has become sadly unbalanced. The second unused residue is of animal origin—the washings of shippons and piggeries and crude sewage. These, if they could be brought into contact with the unused straw, could be used up in compost making.

Ground is being broken in two directions in the salvage of these unused animal wastes. When the washings of piggeries, shippons, and crude sewage from the mains are used to activate straw—loose or baled —excellent compost can be made in three months without any nuisance of any kind. At the moment this pioneering work is being done with hand labour, but when a supply of muck-making machines is available, it will be an easy matter to mechanize this conversion of unused straw into manure.

The second development is taking place in the salvage of sewage. In place of the present-day expensive sewage purification processes, which create wet sludge as an end product, work is in progress to filter off the sludge at the beginning and then to render the effluent harmless by chlorination. In this way a much richer sludge will be obtained. This is being dried and will be put up for sale in 14 and 28 lb. bags, so that the many private gardens and allotments in the urban areas can secure regular supplies of the essential animal wastes for their compost heaps.

Once supplies of dried sewage sludge are available—to supply the essential activator of animal origin—the remaining obstacle to a nation-wide composting campaign in the gardens and allotments of this country will have been removed. Ample vegetable wastes are already available. The composting of small quantities of material is now possible by means of the New Zealand box (p. 219). The only remaining difficulty, soon to be removed, is the supply of animal manure now that the motor-car and the motor-truck have so largely replaced the horse.

The necessary pioneering work in garden composting has already been done. In 1940 a beginning was made in the compost crusade by the County Palatine of Chester Local Medical and Panel Committees, who inaugurated an annual garden competition for the county in which the use of compost was obligatory and artificial manures prohibited. A large number of prizes were offered, as well as three championship cups —one for the best garden or allotment in the county, one for the best rural garden, and the third for the best urban garden. The results are judged by a panel of professional gardeners. On several occasions I have been privileged to see the results, which I felt could not be bettered in any part of England.

Another gardening development has taken place in Westmorland, largely in connection with the activities of Mr. F. C. King, the head gardener at Levens Hall, who has adopted the Indore Process, the merits of which he has explained at a series of evening lectures and in a number of articles published in the *Gardeners' Chronicle* and other journals. Two developments of this work are important. Levens Hall gardens have become a place of pilgrimage for visitors interested in compost gardening; Mr. King has also written two books—*The Compost Gardener* (Titus Wilson & Son Ltd., Kendal, 1943) and *Gardening with Compost* (Faber and Faber Ltd., London, 1944), in which he has emphasized the place of humus in the gardening of to-morrow.

Two developments in nutrition, which have been in progress for some time, are being copied at new centres. At a number of boarding schools the vegetables and fruit consumed by the boys and girls are grown on humus-filled soil.

A second milestone in nutrition has been planted at the Co-operative Wholesale Society's factory at Winsford in Cheshire. Here the potatoes and vegetables used in the canteen meals are grown on fertile soil round the factory with results which have already been described (p. 183). A number of other similar projects are in the making, the results of which will be recorded in the forthcoming issues of the *News-Letter on Compost*.

THE UNITED STATES OF AMERICA

The United States of America are rapidly developing a new type of pioneer in organic farming and gardening—one very different in outlook and methods from the men and women of the covered wagon period who, in their great trek westwards, carved out new farms and created new townships by the simple process of cashing in the soil fertility left by the prairie and the forest. The new trail is leading not towards a new land but to an old and worn out land made new. Success is certain because a sound long term policy is being followed and the snares of the get-rich-quick methods of the manure bag are being avoided. Work is beginning at the source: the compound soil particles on which the stability and permanence of the good earth depend are being re-created and maintained by means of humus. Many areas almost in ruins are being made over.

The disciples of organic farming and gardening are drawn from the entire population. They are either independent pioneers or work as loosely organized groups which have been well described as fellow travellers on a new road. Perhaps the best contemporary account of these adventures in food production is that of Mr. J. I. Rodale in the concluding chapter of his recent book *Pay Dirt* first published in 1945 by the Devin-Adair Company of New York. This work has already reached the fourth edition.

The story thus unfolded proves beyond all doubt that a definite trend towards organic farming and gardening is well under way in America. The new movement is beginning to invade the Experiment Stations where among others Dr. William A. Albrecht of the Missouri College of Agriculture and Professor B. F. Lutman of the University of Vermont are breaking important new ground.

Most of the leading writers on agricultural problems now emphasize the importance of humus. The garden editors of the large metropolitan newspapers constantly mention the importance of adding humus material, through mulches, manure or composts, and one comes across many fine articles on the subject in the largest gardening magazines, farming guides and gardening encyclopedias, publications of the Department of Agriculture, and those of many schools of agriculture.

I can verify all this from my own correspondence. Four years ago when Mr. Rodale asked me to help him in the production of his new monthly—*Organic Gardening*—but few letters reached me from other

American correspondents. Now they are coming in ever increasing numbers by almost every mail. The trickle of four years ago has already passed the snowball stage and threatens to develop into a minor avalanche.

"Hundreds of thousands of agriculturists, amateur and professional, are practicing compost gardening to a small or large extent, experimenting on their own, in many cases, to see whether it is an improvement over the "scientific" way. The fact that enough American farmers are turning to it to warrant farm-machinery manufacturers in making special machinery for handling large scale compost installations testifies to the results the farmers are getting and to the success of their own experimenting." (Rodale)

These promising developments are expected to receive a considerable reinforcement as the veterans of the recent World War are demobilized and return to civil life. Many thousands wish to become farmers and to get their corner of their vast country into good shape, fertile and productive. Two courses are open to these men—the orthodox methods of chemical farming, and the organic methods of their forbears which have stood so well the test of time. A friendly "war in the soil" is bound to result which Mother Earth will decide.

Another important factor in rural life of the United States is the decentralization of industry which has taken place since America entered the war. Factory after factory has been built in the rural areas and in the smaller towns; a great shift of population has taken place; there are hundreds of thousands of people who will never return to the congestion of the swollen cities. The country-side will thus obtain the manifold advantages which flow from the fresh eye on rural questions and from the operations of the reformer and even of the iconoclast. A new America might easily emerge.

Paul Corey in *Buy An Acre* (Dial Press, 1944) has emphasized this new trend in this prediction: 'The country within a radius of from fifty to one hundred miles of our cities will become the New Frontier of America. Ten million tiny homesteads each with an acre or so of ground on which to raise a few chickens and the family's yearly supply of fruits and vegetables, will spring up within commuting distance of factory and business. Congested urban and industrial areas will eventually dissolve over the land.

"This time millions of people will go to the land for homes—an acre, two acres, never more than ten. They will bring this dead land back to life again, build strength again into the soil. The first time we took the land to exploit it; now we are taking the land to save it."

Here we have a promising answer to the ravages of the dust bowls. Catastrophe as usual is bound to compel reform. In days to come the historians of America may even look upon their dust bowl period very much as we in Great Britain now regard the Black Death in the fourteenth century—as a blessing in disguise—as a turning point in history. In England this great epidemic shattered the foundations of the rigid Feudal System and compelled agrarian reform. In the United States the dust bowls may help to replace the thraldom of the profit motive by a still stronger and more permanent directive—real health, contentment and well-being.

There is one simple method of arriving at the reality behind this back to the land movement—by checking the growth of the farming and gardening monthlies. In the case of *Organic Gardening*, with which I have been connected since the beginning, the number of subscribers has grown from zero in May 1942 to some 51,000 in August 1946 in spite of paper shortage and war time publication difficulties.

No less significant is the way the various organized groups such as the School of Living at Suffern, New York, under the leadership of Ralph Borsodi, the many Federal and State agencies, 4-H Clubs, Farm Bureau, Grange, and many rural cooperatives are taking up the task of re-making their country. The National Catholic Rural Life Conference, under the leadership of Monsignor Ligutti, by means of their meetings and their monthly magazine *Land and Home* abound in suggestions for making rural life more satisfactory from the social, spiritual and material aspects. This periodical publishes many practical articles on good, sound farming.

Last but not least is the important Friends of the Land Movement described by Louis Bromfield, one of its leaders, in *Pleasant Valley*. Starting with some sixty members in 1940, it soon passed the 4000 mark, and at once began to issue a quarterly illustrated magazine *The Land* which is always full of valuable and stimulating articles. The membership represents the intelligentsia of the United States and includes bankers, writers, soil-scientists, soil-conservationists, gardeners, artists, County Agents, teachers, farmers and others interested in the re-creation of country communities.

The Friends of the Land Movement is also interesting as showing once again the supreme importance of practical and successful examples so striking as to render all further arguments and discussions unnecessary. One of the most important of these was created by Louis Bromfield himself in Ohio where a group of derelict farms, which he knew in boyhood, were purchased in 1939 and then literally made over and transformed by means of humus. An example of well farmed land for all to see and to copy was soon created. This was achieved on co-operative lines—Mr. Bromfield supplied the inspiration and the capital, a profit-sharing Co-operative Society did the rest. In the concluding pages of *Pleasant Valley* the author writes: "What we want is a new courage and a new race of pioneers as sturdy as the original pioneers but wiser than they—a race of pioneers concerned with the physical, economic and social paradise which this great country could be. . . . These new pioneers will have to be men who understand that the wealth and well-being of every sound nation is founded upon its soil."

That these words are now being written on the land itself will be crystal clear to all those who have studied *Pay Dirt* and the inspiring messages it conveys.

What a contrast this new country provides for that expert-ridden island in the North Sea—Great Britain. The alert Americans are learning about soil fertility by doing because they have not lost that priceless pioneering quality—a willingness to dare. In England the majority of the intelligentsia, enfeebled for the last two generations by de-vitalized food, are still sitting on the fence, vainly waiting till they are furnished with statistical proof that farming and gardening with humus is more profitable than by means of the manure bag and the poison spray, and quite oblivious of the fact that quality can never be weighed or measured and so converted into numbers.

PLATE XIII. COMPOST-MAKING AT CHIPOLI, SOUTHERN RHODESIA

WATERING THE HEAPS

PLATE XIV. THE RAPIER MUCK-SHIFTING CRANE

14

THE RECEPTION OF THE INDORE PROCESS BY THE SCIENTISTS

BEFORE LEAVING India in April 1931 arrangements were made to supply the Indian Central Cotton Committee with a sufficient number of copies of *The Waste Products of Agriculture: Their Utilization as Humus,* so that they could get composting taken up in all the cotton-growing areas without delay. After the book appeared the reviewers all over the world wrote many favourable and even enthusiastic notices, all of which were duly printed. A number of printed slips describing the contents and purpose of the book were then sent to most of the agricultural investigators of the Empire. Ample publicity was in these ways secured. The outcome was interesting and illuminating.

The reception of the Indore Process and its various implications by the experiment station workers engaged on cotton problems proved to be a foretaste of what was to follow. It was, with few exceptions, definitely hostile and even obstructive, largely because the method called in question the soundness of the two main lines of work on cotton —the improvement of the yield and quality of the fibre by plant breeding methods alone, and the control of cotton diseases by direct assault. If the claims of humus and of soil fertility proved to be well founded, it was obvious that this factor would influence the yield much more than a new variety or anything an entomologist or a mycologist could achieve. Besides, both these devices—plant breeding and pest control—would have to wait till the land was got into good heart and maintained in this condition, for the simple reason that any new variety would have to suit a new set of soil conditions, and the inroads of pests might either be prevented or at least reduced by a fertile soil. Further, the current work on chemical fertilizers would have to be postponed till the full effects of a humus-filled soil had been ascertained. The production of compost on a large scale might, therefore, prove to be revolutionary and a positive danger to the structure and perhaps to the very existence of a research organization based on the piecemeal application of the separate sciences to a complex and many-sided biological problem like the production of cotton. Two courses were obviously open to the research workers on cotton: (1) they might save the organization and their own immediate interests by sabotaging the humus idea, or (2) they could give it a square deal and, if it proved successful, could then deal with the new situation from the point of view of the interests of the cotton growers. The vast majority adopted the former course. A few, however, who were engaged in the practical side of cotton growing, took steps to get first-hand experience of humus manufacture and of its effects on the soil and on the cotton crop.

The research workers on most other crops all over the Empire took a similar hostile view and were naturally supported and sustained in their opposition by vested interests like the manufacturers and distributors of artificial manures and poison sprays who were, of course, anxious to preserve and even expand a profitable business. It has been said that even the principle of gravitation would have had a hard row to hoe, had it in any manner stood in the way of the pursuit of profit and the operations of Big Business.

A few examples of the kind of opposition displayed by the laboratory workers and the way in which they were overcome may be quoted. The first of these developed when the tea planters of India and Ceylon began to make compost.

The story of the adoption of the Indore Process by the tea industry has already been told (p. 111) with the exception of an account of the consistent opposition of the tea experiment stations in India and Ceylon to the compost idea. The methods adopted to discredit humus were two.

At first the tea industry was warned that composting was uneconomic and that the game was not worth the candle. Figures were published in Ceylon showing the extra staff needed for the work and the output that could be expected. This put the cost per ton somewhere in the neighbourhood of ten rupees. But a large number of tea gardens were already making first-class compost at less than a fifth of this extravagant estimate, which was based not on actual experience, but on paper calculations. Some of the most important of the tea groups even came to the conclusion that composting cost nothing, as no extra labour or expense was involved because the conversion of wastes into humus was a mere matter of using the existing labour force to the best advantage.

The second line of attack was based on a comparison of the yields of the small plots of the tea experiment station in Assam, where the use of compost and sulphate of ammonia were compared. Results were obtained which appeared to demolish the Indore Process altogether. But these yields, obtained under unnatural conditions on small pocket handkerchiefs of tea, firmly fixed in a strait-jacket as it were, and not provided with shade trees, were flatly contradicted by the large-scale results obtained on many tea gardens. The contest was at its peak when I passed through Calcutta at the end of 1937, when one of the directors of the largest group of tea companies asked me to call upon him. In our conversation reference was made to an abusive article written by one of the advocates of artificials in a periodical devoted to tea which had just appeared in Calcutta, and I was asked if I had seen it. As a matter of fact I had not, but several correspondents had told me of its contents. I was then assured: (1) that no change would be made in the policy of this group which intended to stick to humus, and (2) that orders had already been given that not a single ounce of sulphate of ammonia was to be purchased in future. The controversy was closed by the war which interfered with the import of chemical fertilizers.

These incidents are mentioned to show that the difficulties and delays in getting the law of return adopted in tea were due mainly not to the tea industry, but to advice based on paper calculations and on the yield of small plots growing under unnatural conditions.

One of the best examples in composting I saw in the course of a visit to tea estates in India and Ceylon in 1937–8 was Gandrapara, a garden on the alluvial soils of the Dooars, where excellent management assisted by humus has provided the industry with a safe example to copy. A detailed account of composting on this estate is given in Appendix A (p. 265), from which it will be seen that the yield of tea has gone up by 50 per cent since the time of my visit in 1937. The results obtained illustrate the influence of good farming methods on quality. Gandrapara has moved out of its class and has yielded produce superior to that usual on the soils of this locality.

The next attempt to discredit humus occurred in connection with a project to compost the old hop vines and string on a large garden in Sussex, which had been placed at my disposal by the directorate on condition that I could secure the interest and support of the manager. But the moment this project became known in south-east England it was opposed by the specialists concerned with disease, who argued that my project would mean the destruction of the fine property to which so many years of work had been devoted. To counteract these influences a meeting had to be arranged at East Malling with the specialists of the south-eastern counties and representatives of the Ministry of Agriculture for a discussion on disease: in all some fifty people, almost all hostile to my ideas, took part. I asked the late Professor H. E. Armstrong to accompany me and to observe the proceedings. To give my opponents every chance I prepared a short synopsis of my views and asked the secretary to distribute copies before the meeting. The discussion lasted all day. It was obvious that my specialist opponents, with one or two exceptions, were mere laboratory hermits who had never mastered the art of agriculture, had never grown a crop, and had never taken their own advice about remedies before writing about them. Further, their experience of disease was limited to the conditions of a single island in the North Sea—Great Britain. Only one had visited that cradle of agriculture—the Far East. I had no difficulty in pulverizing the objections these specialists advanced to my thesis that insects and fungi are not the real cause of disease and that pests must be carefully treasured, because they are Nature's censors and our real professors of agriculture. The results of this meeting soon became known. The local opposition to my proposals to convert hop string and hop bine into humus melted away and the project proved to be a great success. Just before the recent war about 10,000 tons of finished humus a year were made on this hop garden from the following raw materials—pulverized town wastes which had to be railed from Southwark to Bodiam, all the wastes of the hops including hop bine and hop string, and every other vegetable and animal waste that could be collected locally. What was interesting was that the all-in cost of preparing and distributing the compost was less than would have been spent on an equivalent dressing of artificials. What was still more important than the saving of money was the beneficial result of the compost on the texture and free working of the heavy soil and on the yield and quality of the hops.

An earlier encounter with the research organization took place at Cambridge towards the end of 1935, when I was invited by the students of the School of Agriculture to address them. I selected as my subject "The Manufacture of Humus by the Indore Method" and distributed printed copies of the gist of my remarks, so that a lively discussion could follow the lecture. Practically the whole of the staff of the Cambridge School of Agriculture attended and an exciting debate followed the

248

lecture. It was an excellent opportunity of trying my medicine on a new dog—in this case, the men engaged in teaching and research. I obtained little or no support for my views from the teachers: if anything, the opposition on the part of the representatives of chemistry, plant breeding, and vegetable pathology was even more pronounced than later at East Malling. The students, however, were not only deeply interested in the subject, but vastly amused at finding their teachers on the defensive and vainly endeavouring to bolster up the tottering pillars supporting their temple. Here again I was amazed by the limited knowledge and experience of the world's agriculture disclosed by this debate. I felt I was dealing with beginners and that some of the arguments put forward could almost be described as the impertinences of ignorance. It was obvious from this meeting that little or no support for organic farming would be obtained from the agricultural colleges and research institutes of Great Britain.

The fourth example of opposition came from the agricultural chemists in the course of the discussion of a paper I read to the Farmers' Club on 1st February 1937 on "The Restoration and Maintenance of Fertility." Representatives of the experiment stations and of the artificial manure industry poured ridicule on my ideas and suggested that they lacked the conventional support of the small plot and the approval of the statisticians. In winding up the debate, I stated that I did not intend to devote any time to a detailed reply to these superficial criticisms, but would shortly have my answer thereto written on the land itself. This was done two years later by the late Sir Bernard Greenwell in one of the most outstanding papers ever read to the Farmers' Club. His large-scale results more than confirmed my paper of two years before. The effect of freshly prepared humus was written by one of the leading agriculturists of the country both on the livestock and on two of his well farmed estates. Although invited to the discussion on Sir Bernard's paper, the representatives of the experiment stations and of the artificial manure interests had no stomach for the fight and did not attend to hear their previous criticisms demolished by the one unanswerable argument—success.

A number of other similar clashes could be quoted, but they would only confirm what has been stated above. These reconnaissances were all carried out for a very obvious purpose—to ascertain the reaction of agricultural teaching and research to the idea that soil fertility is the basis of health in soil, crops, livestock, and mankind. The results showed that in the humus campaign already in progress little assistance could be expected from the official organization. At the same time, it was obvious that nothing need be feared from a body of men engaged on the research side in learning more and more about less and less, and on the teaching side in endeavouring to instil in the rising generation a number of unsound principles based on obsolete methods of investigation. I regretfully came to the conclusion that most of the money devoted by the State to further agriculture by means of the experiment station and the agricultural college has only succeeded in creating an effective bar to all progress and to all new ideas.

The controversy has continued without intermission. Ample space was devoted in a previous chapter ("The Intrusion of Science") to considering the general trend of the scientific researches devoted to agriculture and to analysing where, in my opinion, they have ceased to be effective. The special hostility shown to my own ideas is scarcely surprising and would not be worth special attention here, were it not that the whole vast and expensive machinery of agricultural research is being used to bolster up official authority in Great Britain to deny to the public that freedom of choice which alone can secure progress. Fortified by the findings of Rothamsted and supported by the teachings of the agricultural colleges, the Ministry of Agriculture takes the line that the soil can be kept in good heart by applying still more artificial manures supplemented by the organic matter left by the temporary ley and the dwindling supplies of farmyard manure: the war situation was used to urge this policy on the country.

In thus advocating the temporary ley and in admitting the usefulness of organic matter, my opponents have already travelled a long way from their original point of view. Facts have been too much for them. In refusing to concede the necessity for a well considered national manurial programme based on proper principles, they are still showing themselves to be only tinkerers at the subject—nowhere have I been able to induce them to accept my challenge, take a couple of farms, farm one with artificials, the other on organic principles, and watch the results: nor has any concession been made to my contention that the only satisfactory test of improved pastures, etc., is to *ask the animal*. Neither of these ideas has been received with any favour whatever. Instead, pen is put to paper to prove the efficacy, the benefit, and the absolute need for artificial manures. The latest typical pronouncement is a long reasoned statement by Dr. A. H. Bunting in *Country Life* of 25th February 1944.[1] The statement shows rather exactly the present stage of the controversy about artificial manures and is, therefore, worth analysing.

The gist of Dr. Bunting's case for artificials is given in the two following statements:

1. "The nutrients ordinarily present in the soil are inadequate for continuous intensive production, since the soil is quite unable to supply nutrients at the rate and in the total quantities needed. While it is true that organic manures of various types do contain considerable amounts of these inorganic nutrients, their use cannot supply all that is required on a farm unless the necessary amounts of nitrogen, phosphorus, and potassium are introduced from outside, as in cattle feeding stuffs in certain types of mixed farming. Further, the addition of the complex mixture of nutrients present in such manures gives no possibility of control of the balance of manuring which is so important in practice."

2. "The substances contained in these inorganic fertilizers are, of course, normal constituents of all fertile soils. The importance of the inorganic additions is that they significantly increase the quantities available as distinct from total nutrients, a considerable proportion of which are combined in such a way that they are only slowly available to the plant."

If we analyse these two statements which amount to a heavy indictment of Nature's methods, the argument in favour of artificials falls into three parts: (a) Nature does not supply enough of the inorganic nutrients—they must be supplemented "from outside"; (b) the plant nutrients are not provided by Nature in easily ascertainable quantities and therefore cannot be controlled, and (c) Nature is too slow in her operations to meet present-day needs.

These arguments accuse Nature of being too mean, too inexact, and too slow!

The accusation of meanness lands Dr. Bunting into a difficult position. His suggestion for correcting Nature is a simple one: let us add by our own efforts those extra quantities of food materials which her niggardliness refuses to provide: in this way we shall secure the returns from the soil we desire. These extra quantities are to be brought in "from outside" and he instances feeding stuffs for cattle. But this only amounts to a transfer of natural fertility from one part of the earth to

[1] Reprinted, together with my reply in the same journal on 12th May 1944, in the *News-Letter on Compost* of June 1944.

251

another with no provision for the return of wastes to the land. It is exploitation pure and simple—one of those short-sighted and superficial devices dear to the bandit—in other words, it is the absurdity of folly.

The second argument is that the food materials for the plant supplied by Nature are not provided in easily ascertainable quantities and therefore cannot be "controlled." This is true. But when we attempt to determine these quantities by chemical analysis, the result is failure because, like a census of the population, it only catches the truth at one moment and would have to be endlessly repeated for each small field without pause or intermission if a really exact picture of the state of the soil is to be obtained. Soil analyses have all the disadvantages which follow the application of a static instrument to a dynamic and living system. This being so, the hope that the needs of the plant can be ascertained and then made good is a chimera: the idea that exact weighments of this and that food material can help is to ignore the way Nature acts, to forget the living processes by which the huge reserves in any fertile soil are made available for crops by the work of the soil population. To ignore all this and to talk of a balanced manurial programme is the height of short-sighted folly.

The last argument suggests that Nature is too slow. That accusation is without foundation in all cases where the law of return is faithfully followed. It only holds for worn-out land, where the life of the soil has been starved and the land deprived of its manurial rights. There is no slowness to be seen in the way a well farmed area sets about the growing of a crop. It is an interesting sidelight on Dr. Bunting's allegation that Mr. F. C. King of Levens Hall states that in his experience one of the advantages of well composted land in market gardening operations is that an extra crop per year can be got off the ground: the plants "get away" so much more quickly.

In the course of developing his case Dr. Bunting makes a number of interesting and important concessions. He agrees that the maintenance of the crumb structure of the soil is vital, that the soil needs a constant supply of oxygen, as well as organic matter. He also makes two confessions—that artificials can be abused, and that at Woburn, a branch of Rothamsted, continuous dressings of sulphate of ammonia have been disastrous. His statement, however, leaves much to be desired on the biological processes going on in the soil, on the importance of quality in crop production, and on the power of the crop to resist disease.

Moreover, he has completely ignored the significance of the mycorrhizal association.

In my answer I gave a few examples of the long-term results of artificial manures and cited the case of the sugar industry in Barbados, where of recent years the replacement of organic manure by artificials has led to the virtual collapse of the island through disease and to a decision to re-introduce mixed farming. Another example given was the potato industry of South Lincolnshire, now well on the way to its Tannenberg as a result of the inordinate use of artificials and the reduction in the head of livestock. It is not necessary again to set forth my case—the pages of this book have done so. More especially will a perusal of the examples cited in Appendices A, B, and D completely demolish the case for artificials. Chemicals give increased yields only on infertile or badly farmed land. When these areas are got into first-class condition by means of freshly prepared humus, no artificials are needed. The increased soil population which develops as a result of a humus-filled soil provides the crop with everything it needs.

By 1940 I had come to the conclusion that "the slow poisoning of the life of the soil by artificial manures is one of the greatest calamities that has befallen agriculture and mankind." Nothing has shaken this conviction. It is amazing that the artificial manure interests have not come forward to finance the large-scale trials Lord Teviot and his supporters have pressed for in a recent parliamentary debate. If they are sure of their ground and confident of the final results, what better and cheaper advertisement for artificials could be devised? If the Ministry of Agriculture really believes in its grow-more-food campaign, why did the Minister not move heaven and earth to accept the challenge to his policy of food production and of the present-day organization of agricultural research and teaching? Why not silence *these very tiresome and very persistent advocates of organic farming* once and for all? Refusal to join battle cannot be due to lack of money on the part of the vested interests and of the State. Is the reason for avoiding the fight to be found in another direction altogether—to fear of the verdict of Mother Earth?

PART IV

CONCLUSIONS AND SUGGES-TIONS

15

A FINAL SURVEY

THE NATURAL REACTION to failure is to think again. Perhaps the best known and most vividly expressed example of the ruin which results from choosing the wrong road is that of the Prodigal Son. To-day the realization that there must be something very much amiss somewhere with a civilization which has led us, within twenty years or so, into a second and greater world war, to win which we must pour out all our resources, has produced plan after plan to guide our progress in the future into the paths of sanity and common sense. We are living in an age of planning, in other words in a phase of acute contrition for the blunders of the past.

Why has civilization proved such a disastrous failure? The answer is simple. Our industries, our trade, and our way of life generally have been based first on the exploitation of the earth's surface and then on the oppression of one another—on banditry pure and simple. The inevitable result is now upon us. The unsuccessful bandits are trying to despoil their more successful competitors. The world is divided into two hostile camps: at the root of this vast conflict lies the evil of spoliation which has destroyed the moral integrity of our generation. While this contest marches to its inevitable conclusion, it will not be amiss to draw attention to a forgotten factor which may perhaps help to restore peace and harmony to a tortured world. We must in our future planning pay great attention to food—the product of sun, soil, plant, and livestock—in other words, to farming and gardening.

What is the place of farming and gardening in human affairs? We can best answer this question if we bear in mind what are the essentials needed by mankind. They are five in number and in order of importance they are: air, water, *food,* warmth, and shelter. Without a supply of air life lasts but a few minutes; without water, only a few days; without food it is only possible for the human body to exist on compensation for a few weeks. We can, to a large extent, control the warmth factor by making the fullest use of our own animal heat. The question of shelter, often described as the housing problem and to which most attention is now being paid by the planners, is the least important of the Big Five, which must always be at the basis of all our future schemes.

Our food is produced for the most part by farmers and gardeners. It has been sadly neglected in the past, as will be clear to anyone who studies this book and its many implications. The essential things about food are three: (1) it must be grown in fertile soil, that is to say in soil well supplied with freshly prepared, high quality humus; (2) it must be fresh; (3) its cost must be stabilized in such a manner as to put an end to the constant fluctuations and steady rise in prices. All these things are possible once we increase the efficiency of the earth's green carpet—the machinery furnished by Nature for producing food. *The sun provides the energy for running this mechanism, so our power problem has been solved for us.* The sole food producing machine is the green leaf. This, again, is the gift of Providence. Mankind can increase the efficiency and output of this green carpet at least threefold by (1) the restoration and maintenance of the fertility of the soil on which it rests and (2) by providing varieties of crops which make the most of the sun's rays and the improved soil conditions. The former can be achieved by converting into humus the vast stores of vegetable and animal residues now largely running to waste: the latter by modern plant-breeding methods. Once we do this, all goes well. The roots are provided with a favourable climate and ample living space. The yield and quality of the produce go up by leaps and bounds: the danger of any shortage of food in the world disappears: the problem of price regulation is automatically solved.

How can the increased efficiency of the green carpet help in stabilizing prices? In a very simple way. Every article we purchase, every amenity we enjoy—such as those connected with defence, transport, the heating and lighting of buildings, the various services connected with news and so forth—all depend on food, because the multitudes of men and women who provide these things for us do not grow their own nutriment: it is grown for them: it is even brought to their tables: all this has to be paid for. The cost of food, therefore, enters not only into what we ourselves consume, but into everything we enjoy individually or in common. Once this food is as abundant as possible, we obviously reduce its cost. The efficiency of the earth's green carpet is, therefore, a fundamental question. Any discussions about price regulation, tariffs, exports and imports, gold standards, and so forth can only be superficial unless they go down to the foundations of our world—the smooth working of the green carpet which manufactures the food, on the cost of which all other prices must depend. There is no other foundation for these discussions on economics. It follows, therefore, that we must take careful note of the basic principles underlying our food supplies. Once these are as abundant as Nature intended they should be, they will be as cheap as it is possible to make them. The regulation and stabilization of future prices then follows. After that, all we have to see to is to prevent anybody or any nation trying to interfere with the free interchange of the direct and indirect products of solar energy from one part of the world to another, because the various regions of this planet differ greatly in the materials they can best provide. Our supplies of sugar, for example, can most cheaply be obtained from the sugar-cane, a tropical or sub-tropical crop: our clothing should come not from processed wood, but from the wool of sheep, an animal which thrives best in rather dry, temperate regions. Our future trading arrangements must, therefore, be based on two things: (1) the full utilization of the sun, and (2) the free interchange of the products of sunlight.

We can check our food production methods by means of Nature's censors—the diseases of crops and livestock. Provided we prepare the soil for its manurial rights by suitable cultivation and subsoiling, and then faithfully comply with Nature's great law of return by seeing to it that all available vegetable, animal, and human wastes are converted into humus in suitable heaps or pits outside the land or in the soil itself by the processes of sheet-composting, we shall soon find that many striking things will begin to happen. The yield and quality will rapidly improve: the crops will be able to resist the onslaughts of parasites: *well-being and contentment,* as well as the power to vanquish disease, *will be passed on to the livestock which consume them:* the varieties of crops cultivated will not run out, but will preserve their power of reproduction for a very long time.

The objection to composting on the average farm or market garden on the score of the dearness and scarcity of labour is being removed by the mechanization of the manure heap. Several machines have already been devised which will assemble the compost heaps, turn them, and load the finished humus on to suitable manure distributors. With the help of one of these machines the cost per ton has already been reduced to less than a quarter. This suggests that mechanized organic farming and gardening is certain to prove much cheaper than the methods now in use, where the manurial rights of the soil and of the crop are being largely evaded by substitutes in the shape of artificial manures. Large-scale results coming in a growing torrent from all over the world show that the ephemeral methods of manuring, by means of chemicals and the resulting *survival of the weakly plant bolstered up by poison sprays,* are bound to be swept into the oblivion which they merit.

The disciples of Rothamsted, which include the Ministry of Agriculture, the experiment stations, and the agricultural colleges, have combined forces with the vested interests concerned with the production and sale of chemical fertilizer and protective poisons for the crop to deflect the onward march of organic farming and gardening. The war in the soil is now in full swing. The first battle has just come to an end in South Africa: it lasted some ten years: it has ended by the conversion of South Africa to humus: the protagonists of chemical fertilizer have taken the count. Two factors which have contributed to this result must be mentioned: (1) the spate of ridicule and abuse which the representatives of chemical farming first poured on humus, and (2) the failure of the artificial manure interests to take up land alongside the pioneers of organic farming and show the country what their wares could accomplish. They unconsciously gave organic farming an excellent advertisement: they had no stomach for the real fight because they feared that the verdict of Mother Earth on their pretensions would be adverse. In Great Britain the same fatal blunders are being made: abusive articles in the press are being relied on rather than *a fight to a finish on the land itself.*

The power to resist diseases, which organic farming and gardening confer on the plant and on the animal, is duly passed on to mankind. The evidence in favour of this view is rapidly growing. When examples without end are available, showing how most of the malnutrition, indisposition, and actual disease from which the population now suffers can be replaced by robust health by merely living on the fresh produce of fertile soil, it will be a simple matter in any democratic country for the people to insist on their birthright—fresh food from fertile soil—for themselves and for their children. The various bodies which now stand in the way of progress will be rapidly eliminated once their interests come in conflict with those of the electorate.

There appears to be a simple principle which underlies the vast accumulation of disease which now afflicts the world. This principle operates in the soil, the crop, the animal, and ourselves. *The power of all these four to resist disease appears to be bound up with the circulation of properly synthesized protein in Nature. The proteins are the agencies which confer immunity on plant, animal, and man.* We must, therefore, first study the nitrogen cycle between soil and crop, and then see to it that the green leaf can build up proteins of the right type. Then there will be little disease in soil or crop or livestock, and the foundations of *the preventive medicine of to-morrow* will be laid. Properly synthesized vegetable protein will confer on the animal and then on mankind the power to overcome infection and to reduce disease to what in the future is certain to be *its normal insignificance.* We shall then discover that the present vast and expensive fabric of social services has been built on the basis of malnutrition and inefficiency. Their foundations will have to be recast to suit a population in good health. The reformed services will obviously cost much less than they do now. A new system of preventive medicine and of medical training will at the same time arise. The physician of to-morrow will study mankind in relation to his environment, will prevent disease at the source, and will cease to confine himself to the temporary alleviation of the miseries resulting from malnutrition.

One of the great tasks before the world has been outlined in this book. It is to *found our civilization on a fresh basis—on the full utilization of the earth's green carpet.* This will provide the food we need: it will prevent much present-day disease at the source and at the same time confer robust health and contentment on the population: it will do much to put an end automatically to the remnants of *this age of banditry now coming to a disastrous close.* Does mankind possess the understanding to grasp the possibilities which this simple truth unfolds? If it does and if it has the audacity and the courage to tread the new road, then civilization will take a step forward and the Solar Age will replace this era of rapacity which is already entering into its twilight.

APPENDICES

APPENDIX A

PROGRESS MADE ON A TEA ESTATE IN NORTH BENGAL

BY J. C. WATSON

GANDRAPARA TEA ESTATE is situated on low rice-growing land south of the Himalayas and in a district which was commonly thought to be incapable of producing teas of a quality equal to those of estates situated on the Red Bank soil. The estate covers 2,796 acres, of which 1,242 acres are under tea; there are also ten acres of seed-bearing bushes. Paddy or rice land is available for the labour force, allotments for growing soya bean, vegetables, and so forth, and *shajana* trees grow in all the labourer's *barees* or garden patches. Everything possible is being done to improve and maintain the nutrition and health of the labour force and also of the labour force of to-morrow—the children. Large sums are being well spent by the Company to maintain a healthy and contented labour force which is one of the finest assets of an estate. I have had the privilege of managing this estate for thirty years and not only has the labour force been contented, happy, and healthy, but the land itself has also improved.

There are resident on the estate a population of 2,756 souls, as well as *two and a half million tea bushes,* all to be maintained in a state of health. The tea plant requires a fertile soil and this means healthy crops, healthy animals, and last, but not least, healthy human beings. The following facts tell their own story: in the five years previous to the intensive application of humus the estate averaged yearly 795,801 lb. of tea or 5.09 oz. of tea per bush; since 1939 22,000 tons of humus, made in a central factory on the Indore method advocated by Sir Albert Howard, have been applied to the land and the yields during 1939–43 averaged 1,240,800 lb. of tea yearly or 7.94 oz. per bush.

It is undeniable that this humus is the storehouse of surplus water which is given back to the plant in dry periods. In this part of India droughts are sometimes very severe; in the period from October to April less than one and a half inches of rain has been registered, but the condition and health of the bushes compared with those estates treated wholly with artificial manures is remarkable. The art of cultivation consists in getting the humus to a depth in the soil where the moisture does not evaporate. The higher the fertility of the soil, the better the class of crop grown on it and the less are the effects of dry periods on the crops. The drainage system where heavy rainfall is experienced—as much as 125 inches between May and September—has to be in thorough working order to keep the soil in good heart, and there has yet to be found any better method of replacing the losses in the soil year after year than by heavy applications of organic matter. If the tea bushes receive a check, they are immediately liable to disease.

It was, therefore, essential that before starting on heavy applications of humus the drainage system be put and kept in good working order, also good shade trees were established giving a heavy leaf fall. There is no substitute for organic matter or humus in the soil. It is interesting to note that in 1943 a severe hail storm stripped the bushes and did damage estimated at 96,000 lb. of tea, but, after resting, the bushes had the stamina to ensure a rapid return to normal and a record crop was harvested.

In 1934 the manufacture of humus on a small scale was instituted according to the Indore method advocated by Sir Albert Howard. The humus is manufactured from the waste products of tea estates. All available vegetable matter of every description, such as *Ageratum*, weeds, thatch, leaves, and so forth, is carefully collected and stacked, put into pits in layers, sprinkled with urinated earth to which a handful of wood ashes has been added, and then covered with a layer of broken up dung and soiled bedding, after which the contents are watered with a fine spray—not too much water, but well moistened. This charging process is continued till the pit is full to a depth of from three to four feet, each layer being watered with a fine spray as before (Plates VIII and IX).

To do all this it was found necessary to have a central factory, so that the work could be controlled and the cost kept as low as possible. Details of the central factory which was erected are given in the plan (Plate XI). There are 41 pits each 31 x 15 x 3 feet deep; the roofs over these pits are 33 x 17 feet, space between sheds 12 feet, and between lines of sheds 30 feet, and between sheds and fencing 30 feet. This allows materials to be carted direct to the pits and also leaves room for finished material. Water has been laid on—a two-inch pipe with one-inch standards and hydrants 54 feet part, allowing the hose to reach all pits. A fine spreader-jet is used; rain-sprinklers are also employed with a fine spray. The communal cowsheds are situated adjacent to the humus factory and are 50 x 15 feet each, and can accommodate 200 head of cattle. The enclosure, 173 x 57 feet, is also used to provide outside sleeping accommodation. There is a water trough, 11 feet 6 inches by 3 feet wide, to provide water for the animals at all times. The living houses of the cow herds are near to the site. An office, store, and chowkidar's house are in the factory enclosure. The main cart-road to the lines runs parallel with the enclosure and during the cold weather all traffic to and from the lines passes over this road, where material that requires to be broken down is laid and changed daily as required. Water for the factory has a good head and is plentiful, the main cock for the supply being controlled from the office on the site. All pits are numbered, and records of material used in each pit are kept, including cost; turning dates and costs, temperatures, watering, and lifting, etc., are kept in detail, Weighments are only taken when the humus is applied, so as to ascertain tasks and tons per acre of application to mature tea, nurseries, tung *barees*, seed-bearing bushes, or weak plants.

The communal cowsheds and enclosure are bedded with jungle and this is removed as required for the charging of the pits.

I have tried out pits with brick vents, but I consider that a few hollow bamboos placed in the pits give a better aeration, and these vents make it possible to increase the output per pit, as the fermenting mass can be made four to five feet deep.

Much care has to be taken at the charging of the pits so that no trampling takes place and a large board across the pits avoids the possibility of coolies pressing down the material when charging. At the first turn all woody material that has not broken down by carts passing over it is chopped by a sharp hoe, thus ensuring that full fermentation may act, and fungous growth is general.

With the arrangement of the humus factory compost can be made at any time of the year, the normal process taking about three months. With the central factory much better supervision can be given and a better class of humus is made. That made outside and alongside the raw material and left for the rains to break down acts quite well, but the finished product is not nearly so good. It therefore pays to cut and wither the material and transport it to the central factory as far as possible.

In the cold weather a great deal of sheet-composting is being done. After pruning, the humus is applied at the rate of seven to ten tons to the acre and hoed in with the prunings, the bulk of which varies. In this way excellent results have been obtained. The pits become small composting chambers; the roots of the tea bushes soon invade the pits, and results speak for themselves.

On many gardens the supply of available cow-dung and green material is nothing like enough for requirements. Many agriculturists try to make up the shortage by such expedients as the hoeing in of green crops and the use of shade trees or any decaying vegetable matter that may be obtainable; on practically all gardens some use is made of all forms of organic materials and fertility is kept up by these means. It is significant to note that for many years now manufacturers who specialize in compound manures usually make a range of special fertilizers that contain an appreciable percentage of humus. The importance of supplying soils with the humus they need is obvious. I have not space to consider the important question of facilitating the work of the soil bacteria, but it has to be acknowledged that a supply of available humus is essential to their well-being and beneficial activities. Without the beneficial soil bacteria there could be no growth and it follows that, however correctly we may use chemical fertilizers according to some theoretical standard, if there is not in the soil a supply of available humus, there will be disappointing crops, weak bushes, blighted and diseased frames. It would, moreover, be to the good if every means whereby humus could be supplied to the soil in a practical and economical way could receive the sympathetic attention of those who, at the present time, mould agricultural opinion.

To the above must be added the aeration of the soil by shade and drainage. I am afraid many planters and estates do not fully understand this most important operation in the cultivation of the tea bush. To maintain fertility we must have good drainage, shade trees, and tillage of various descriptions to kill weeds. The best areas are the cleanest, and not only do they secure bigger crops and higher quality, but they have nothing to waste.

Humus is essential: artificials are a tonic, but humus is a food. It is not difficult to understand that the use of artificials in feeding the plant direct sidetracks a portion of Nature's essential round. Artificial stimulus, applied year after year and at the same times, must inevitably breed evils, the full extent of which are yet but dimly seen. The time may come when yield will depend entirely on quality, but quality can never under any circumstances depend upon yield. Factory-made manure is the weak link in the chain of agricultural economies. Humus is the real food of the soil and the crop; it leads to and maintains larger crops and improved quality.

For the past five years no chemical manures or sprays for the control of disease and pests have been used. The return to the soil of all organic waste in a natural cycle is considered by many scientists to be the means of obtaining the best teas and of resisting pests and disease. The tea bush requires nutrition, and Sir Albert Howard not only wants to increase the quality of human food, but, in order that it may be of proper standard, he wants to improve the quality of plant food. That is to say, he considers the fundamental problem is the improvement of the soil itself, making it healthy and fertile. "A fertile soil," he says, "rich in humus, needs nothing more in the way of manure: the crop requires no protection from pests: it looks after itself. . . ." It is interesting to note that plant diseases are the consequence of infertility, so that the rational method of dealing with such problems is not to destroy the agent by means of insecticides and fungicides, but to bring the soil back into a condition of real fertility in the first instance, and then to devise the best methods to suit local conditions.

Gandrapara Tea Estates,
Banarhat P.O., Dooars.
10th August 1944.

APPENDIX B

COMPOST MAKING IN RHODESIA

BY J. M. MOUBRAY

In 1939, when I last wrote a few notes for *An Agricultural Testament*, compost making in Rhodesia was in its infancy. Now it has become general. The usual procedure now adopted is to break down the vegetable wastes by spreading them in stock-yards or pens. Here they absorb and get well mixed with the animal wastes both solid and liquid, and are then removed to the compost heaps. In this part of the country growth is very rank. When tall grass and reeds were moved straight to the compost heap, the stems took a considerable time to break down, but by being first trampled down the stems are broken and the fungi and bacteria are then able to attack both from the inside and outside at once.

In the five years that have passed since 1939 little change in procedure has taken place with the exception of passing all raw material through the stock-yards. I still build the heaps some fifteen feet wide and three feet high and up to any length (Plate XII). Two turnings are sufficient and at the end of three months the breakdown is complete. In the dry weather, if the heaps are fairly moist when built, a good wetting with the hose-pipe each time the heaps are turned is sufficient (Plate XIII). Material from the outside of the heap is always turned inside. I cut a good deal more hay than I used to do and if some of this is a bit coarse or gets a wetting, it does not matter, as what the cattle do not eat goes to the compost heap. Our veldt is improving with mowing, as when the coarse grasses are kept down and in check the finer and more valuable grasses get a better chance to develop.

We are learning that under conditions in many parts of Mashonaland nitrification is very rapid. Under favourable moisture conditions a green crop ploughed in leaves little visible organic matter at the end of three months. Partly for this reason, if the compost is not quite broken down when applied to land for crops like maize, we get better results.

The nitrogen content of compost has been found to be quite stable. I have found the loss of nitrogen in a heap which has stood for some months in the dry weather to be negligible. Mr. van Vuren, who has done so much in the Union of South Africa for municipal compost, has found much the same to happen with him. I now spread out some of my compost in a thin layer. In the hot sun this gets quite dry in a day or two. I then grind it in a hammer mill, sack it, and it can be kept in such a manner for an indefinite period. In this way it has probably lost some 40 per cent of its moisture content and is so correspondingly richer in humus. If, instead of broadcasting rough compost, a cupful of the ground material is applied round the plant in the field for such crops as tobacco or tomatoes, a considerable economy is effected.

I add ground raw rock phosphate to all my compost heaps. It is probable that some of the inorganic phosphorus is changed during the fermentation into organic forms. If this is so, and some of the best American opinion considers such a change takes place, it is all to the good, as in its organic forms phospohorus is not locked up and so made unavailable to the plant, as it does not combine with iron and alumina.

As regards cost of making compost, assuming that bedding of some sort has to be provided for the stock-yards and that the work of cutting and carting such bedding is debited against the stock account, then I think most farmers in this part of the world will agree that a sum of 1s. or 2s. per ton will cover the cost of compost making. That is, of course, apart from the cost of raw rock phosphate or similar material added to the heap.

The effect of compost on fruits, vegetables, and field crops in Rhodesia is now so well known that further propaganda is unnecessary. A neighbouring farmer, to give one example, used it on bananas and found that in two seasons he not only doubled the size of the bananas, but doubled the numbers held in the bunch besides greatly improving their flavour.

The trouble now is that we cannot make enough compost. With labour becoming more difficult various mechanical devices for handling and turning compost are coming into use. An ordinary dam scoop with the bottom elongated by means of steel fingers acts very well in moving the material to make the compost out of the stock-yards, and in turning the heap itself. I find nothing to beat hand labour. Once a native gets into the work he will do a large tonnage per day and nothing mixes the material so well as hand labour. If the material is fairly damp and requires little wetting, then two natives, working side by side, keep pace with the hose-pipe; but if it is very dry, then one turner only is used, so that more water can be applied as it is thrown over.

In Rhodesia compost has been found to control the parasitic plant, witchweed (*Striga lutea*), which attaches itself to the roots of the maize. Witchweed used to be a major problem, but on my farm it is now negligible.

It is now being accepted that, in the same way, good applications of compost will eliminate eelworm. This pest had begun to assume very serious proportions in tobacco lands to such an extent that infested lands were considered unsuitable for further tobacco crops.

Organic farming is coming more and more to the fore in Rhodesia. It is at last being recognized that many of our troubles were due to lack of humus in the soil.

Green cropping is taking a larger and larger part in the rotation and the chief plant used is the legume, *san* hemp (*Crotalaria juncea*), this on good soil grows eight to nine feet high and ploughs in very well with a tractor-drawn disc plough. If a light dressing of compost, containing a good proportion of animal wastes, is added to the soil for such a leguminous green crop, more seed is formed. This may be due to the plant growth substances which originate in the animal and perhaps further supplies are formed during decomposition in the compost heap.

Compost and, in fact, all organic matter appears to have considerable effect on the mycorrhizal growth. I speak now, in particular, of the orange tree, of which I have many thousands growing on this farm, Chipoli. If the hair-like feeder roots of a healthy tree are carefully exposed, they will be seen to be covered with a mould-like growth, but if the same is done to an unhealthy orange tree, showing signs of decline, then this is found to be absent.

And now to give what I consider to be one of the best examples of chemicals versus organics. There are in this Mazoe valley two orange groves, both of considerable extent, planted about a quarter of a century ago, of the same variety of orange, the Valencia Late. The trees grow on the same type of good red soil, well drained and irrigated in the dry weather. In fact, conditions are about as similar as they could be. One grove has been fed almost exclusively on artificials—superphosphate, muriate and sulphate of potash, nitrate of soda, and sulphate of ammonia, this last in large proportion. Cultivation is more or less clean, little weed growth being allowed and little or no organic matter applied. The trees in this grove are now practically finished; new growth has all but ceased. The trees are full of dead wood and the crop of oranges they now carry is sub-economic. The foliage is sparse and of an unhealthy colour. In the other grove the only fertilizer used has been raw rock phosphate and bone, but since the start of the war bone has been unprocurable. A heavy green crop of legume is grown during each rainy season, this is broken down and disked in, and the soil is covered with old grass, trash of all kinds, ground nut haulms, and so forth. Irrigation is then applied, when a rank growth of grass and weeds of all sorts comes up through the mulch. This is eaten off *in situ* by cattle and sheep whose droppings fall on the vegetable wastes. With the advent of the rains what remains on the surface and has not been assimilated by the soil bacteria is disked in and the cover crop is at once planted. One has only to look at the trees to see that they thrive. They carry heavy crops of good-quality fruit, the foliage is a dark green, the trees carry no dead wood, and regularly put out a thick new growth.

This example of two treatments is, I think, almost unique. It shows the culminative effect of a treatment of chemicals and of organics over some twenty-five years. These groves are open to inspection by any and all, and the owners will confirm the treatment under which they are grown.

What is the explanation? The accumulation of the sulphate ion in the chemically treated grove must be considerable. Is it this that has prevented the mycorrhizal connection functioning, or is it the lack of humus, or both that have been slowly killing the trees?

One fact emerges and on this there need be no further argument—the orange tree under the conditions described will not thrive for any lengthy period on chemical food alone, but it will do so on organic food. Whether the healthy trees would have been more healthy still if chemicals had been added to the organics, or whether the sulphate ion would have been too much for the mycorrhiza I cannot tell. To prove this conclusively would require another quarter of a century and that is a good deal more than is left to me.

Chipoli,
Shamva, Southern Rhodesia.
27th July 1944.

APPENDIX C

THE UTILIZATION OF MUNICIPAL WASTES IN SOUTH AFRICA

BY J. P. J. VAN VUREN, M.SC. (AGRIC.)

*Professional Officer (Extension) and Co-ordinating Officer,
Municipal Compost Scheme*

LITTLE WAS it realized in August 1939, when the first sod was turned for the excavation of an experimental compost pit somewhere on the boundary of the Ficksburg town commonage, that history was being made. Had this been known at the time, the criticism and prejudice which had to be faced and fought for so many months to come would then have mattered even less than they did.

Up to that time hardly anybody in the country had shown any practical interest in the conversion of otherwise useless and obnoxious products such as garbage, night soil, etc., from urban areas. My own knowledge of this subject was limited to a mere study of the results obtained overseas by men like Howard, Wad, Watson, Jackson, and others. I felt thoroughly convinced, however, that this method could be successfully employed in South Africa if only one municipality could be persuaded to co-operate in the initial experiment or demonstration.

About the time referred to above the author was transferred to Ficksburg in the Orange Free State, a small town with a population of scarcely 3,000 Europeans and situated on the border of the Basutoland Native Territory. On my arrival in my new sphere of activity the matter was discussed with the local health inspector, who at once declared himself willing to co-operate in the laying down of an experiment.

At first a small-scale trial was conducted, well away from the public eye and almost in secret. No funds were available. Ordinary trenches 12 x 8 x 2 feet deep, were dug in the soil and old pieces of scrap corrugated iron were cut, perforated, and used over drainage channels in the floor of the pit. Dry refuse straight from the tipping wagons was dumped in the pit and levelled into a layer about fifteen inches deep. On the top of this came night soil, followed up with refuse and so on until in about three days' time the pit was filled. Right from the outset problems and numerous difficulties were encountered. Owing to poor drainage and the absence of aeration facilities the contents of the pit became a cold, sloppy, reeking mass. Consequently none of the labourers, whose customary task it was to dig trenches for the usual burial of night soil, could be persuaded to do the necessary turning over of the contents—and they could hardly be blamed for refusing. The sides of the pit caved in during subsequent rains and myriads of flies issued from the sodden mass. Fortunately very few outsiders knew at that time what was happening, otherwise our experiment might have ended in court.

However, where there's a will there's a way. Our mistakes were gradually rectified and one after the other our problems disappeared until the stage was reached when an invitation to certain members of the Council could be risked. Their visit had the desired effect and a small sum of money was granted for the erection of proper brick and cement installations. In these new pits, erected according to Watson's Tollygunge plans as described by Sir Albert Howard in his pamphlet *The Manufacture of Humus from the Wastes of the Town and the Village,* excellent results were quickly obtained. Temperatures started to climb to surprisingly high levels. Fly-breeding was prevented by these high temperatures and within four weeks the final product was a dark crumbly mass with no unpleasant odour and without any trace of its original constituents.

At this stage the local authority became convinced of the practicability of the composting process and it at once decided that this "modern" method of urban refuse disposal should receive more sympathy and support. It was consequently decided that a more convenient site should be selected and the scheme extended to include at least fifteen pits instead of only two as was the case up to that time. The ultimate site selected was situated only half the distance from town of that where night soil had been regularly buried for over fifty years, the period of Ficksburg's actual existence. The Council at once realized that a considerable saving on transport would result quite apart from the fact that the final product might be sold, thereby increasing the revenue of the town and consequently reducing the cost of refuse disposal.

Based on the valuable experience gained during the experimental stage of the scheme, the new pits were built accordingly. Certain modifications were introduced and these included the following: an increase in the number of cross channels in the floor from two to seven; vertical side walls instead of sloping ones; an increase in the length and width of the pits and also in the gradient from one end of the pit to the other, the latter to facilitate the handling and distribution of night soil. In addition a shed was erected to protect the final product against wind and weather.

From then on practically all night soil and refuse from the urban area was removed to this new site, where it was turned into compost at the rate of about 100 to 150 cubic yards per month. The refuse included, more or less, the following: the contents of garbage bins minus the coarse pieces of unburnt coal and other refractory material which are screened out on arrival at the site; weeds; grasses; hedge clippings; stable manure; papers; rags; abattoir refuse such as paunch contents, portions of the intestines, rejected meat or organs, blood, etc. (horns, hoofs, and bones were also collected but sold directly to bonemeal and fertilizer factories); sawdust; street sweepings, fallen leaves, etc. No longer were these constituents allowed to be dumped somewhere along the approaches of the town where rats and flies could breed unmolested. Instead, they were henceforth carted to one depot and there rendered harmless by being properly composted.

This, briefly, is the history of composting at Ficksburg. It may, however, be stated unhesitatingly that without the undaunted assistance of Mr. H. G. Williams, the Health Inspector at the time, as well as the sympathetic co-operation of the Ficksburg Town Council (through the medium of their energetic and capable Town Clerk), it is doubtful whether the scheme would ever have developed into the great success it is to-day. Without their valuable assistance Ficksburg would just have remained an ordinary Free State town, whereas to-day it is well known, not only in this country but overseas as well, as one of the pioneers in the direction of urban waste utilization.

No sooner were the first articles published in connection with the preliminary experiments at Ficksburg than inquiries started to pour in from various parts of the Union of South Africa, Rhodesia, Belgian Congo, and East Africa. At the same time a host of visitors were received and shown over the scheme at Ficksburg. According to the correspondence received, most of the urban authorities seemed to be faced with the same problems and difficulties of refuse disposal. This process of composting and getting rid of such material sounded to them like an answer to their prayers with the result that they were anxious to obtain details

PLATE XV. THE RAPIER MUCK-SHIFTING CRANE

in regard to the process as quickly as possible. It did not then take long for the process to become adopted by various centres in southern Africa.

Owing to the fact that South African soils are generally deficient in phosphates, this country is dependent for her phosphate supplies from overseas. When war broke out, shipping facilities were reserved for the importation of essential war supplies. Imports, as far as this commodity was concerned, dropped to about 50 per cent of the pre-war supplies. At the same time there was an increased demand for food at this stage when farmers could obtain only half their normal requirements of fertilizers. As a result of this shortage all possible avenues of obtaining fertilizing material in the country were explored. Farmers were encouraged to give more attention to neglected manure heaps on their farms and to conserve and use this valuable material more extensively than in the past. In addition, farm composting methods were demonstrated and encouraged. Bat manure and phosphate deposits were explored and in some cases made available to farmers in the crop-producing areas. At the same time a huge trade developed in sheep and goat manure from the Karoo, South Africa's principal small-stock area, ultimately reaching such proportions that it was feared that the supplies would not outlast the war. In spite of the exploitation of all these sources of supply, it was still felt that production might suffer from a shortage of the necessary fertilizer material. It was the imminence of this possibility which caused greater attention to be given to the preparation of urban compost. If vegetable and fruit farmers, it was thought, could be encouraged to use urban compost more extensively, then more of the mineral fertilizers would be available for use in the production of grain crops such as maize and wheat. The possibilities of urban compost fulfilling part of this programme were investigated by a special Departmental Compost Committee on whose advice the Department of Agriculture and Forestry decided to institute an urban compost campaign on a national basis, the author being appointed co-ordinating officer for the scheme for the duration of the war. To assist him, six other officials, stationed throughout the four provinces of the Union, were also designated for this work. The duty of these officers was mainly to visit each urban centre in their respective areas and to encourage the adoption of the composting process.

For the purpose of gaining first-hand knowledge and experience of the process, these regional officers met at Ficksburg in August 1942, immediately after the decision to inaugurate this scheme. Apart from studying the method in its various aspects, these officers in conjunction with the co-ordinating officer drew up a programme of action so as to ensure the co-ordination of advice and policy. This programme included the following:

1. The co-ordinating officer was to draw up a specified plan of the pits, as well as a pamphlet describing the Ficksburg composting process in detail, and to issue these to the regional officers for distribution to municipalities in their areas.

2. Until such time as the co-ordinating officer was available to accompany each regional officer in turn through his area, these officers were to leave no stone unturned in so far as preliminary propaganda in this connection was concerned. At about this time the annual Municipal Conferences were to be held in the different provinces and they had to be addressed on the subject. Articles were to be written and published in local papers, etc.

3. By this time there was at least one centre in each of the six areas where the process had been adopted already. At such centres regional officers were to organize two-day short courses for representatives of neighbouring towns. On these occasions practical demonstrations and lectures were to be given so as to make such representatives as thoroughly conversant with the process as possible.

4. Radio talks and articles for the daily press were to be drawn up or circularized.

5. Certain aspects of the process warranted further investigation and in particular the co-ordinating officer was to be responsible for the carrying out of this work at Ficksburg.

This, briefly, was the programme drawn up at the Ficksburg Conference in August 1942, and within six months practically the whole of the Union with its 300 municipalities and health boards was covered. It was soon found that almost all centres were confronted with the same difficulties and problems. Literally mountains of "waste" were encountered at many places. These had accumulated over many years in some cases and it was not uncommon to see, lying in sight of these huge dumps, lands where the soil had been worn down to a condition of total impoverishment. That this has been and is still going on in many centres of the Union even to-day is incontrovertible proof of the naked truth of the late Professor King's words, "Man is the most extravagant accelerator of waste the world has ever endured." Fortunately South Africa is a country of vast open spaces, otherwise dumping sites might have become so limited that many of these dumps of fertility would have had to disappear in clouds of smoke, instead of still being there to-day in a state in which their fertility is still partly recoverable if only urban authorities can be persuaded to render such material marketable in the form of refuse-dump screenings and compost. These "humus mines" as Sir Albert Howard calls them, are in many instances ready for immediate use on the land and could contribute materially to a reduction in the existing shortage of fertilizers.

After two years since the inauguration of the compost scheme, the position in regard to its adoption in South Africa is as follows:

In the various provinces the following towns and cities have adopted the urban composting process:

Northern Transvaal: Nylstroom, Potgietersrust, Pietersburg, Messina, Hercules, Zeerust, and Pretoria (Indore compost).

Southern Transvaal: Potchefstroom, Klerksdorp, Ermelo, Brakpan, Heidelberg, Volksrust, Boksburg, Randfontein. Lichtenburg, Alberton and Johannesburg, Roodepoort, Maraisburg (Indore compost).

Orange Free State: Ficksburg, Ladybrand, Clocolan, Bethlehem, Harrismith, Vrede, Reitz, Heilbron, Parys, Kroonstad, Kopjes and Bloemfontein, Kimberley (Indore compost).

Natal: Matatiele, Glencoe, Stanger, Dannhauser, Vryheid, Howick, Margate, Darnall, Bergville and Durban, Pietermaritzburg (Indore compost).

Karoo and Eastern Cape Province: Aliwal North, Elliot, Fort Beaufort, Graaff-Reinet, Kirkwood, Kingwilliamstown, Prince Albert,

Area	Schemes in		Average Annual Production in Cubic Yards	Total Production to date in Cubic Yards
	Production	Course of Construction		
Northern Transvaal	7	2	5,000	6,300
Southern Transvaal	12	1	13,500	18,600
Orange Free State	13	1	13,250	19,250
Natal	11	7	12,000	15,500
Karoo and Eastern Cape Province	16	6	5,600	12,100
Western Cape Province	22	6	21,000	27,000
Total	81	20	70,350	98,750

279

Queenstown, Umtata, Walmer, Cradock, Dordrecht, Oudtshoorn, Uitenhage, Humansdorp and Beaufort West (Indore compost).

Western Cape Province: George, Parow, Goodwood, Wolseley, Stellenbosch, Mossel Bay, Bellville, Swellendam, Vredenburg, Heidelburg, Robertson, Tulbagh, Capetown, Rivier-Zonder-End, Franschhoek, Ceres, Worcester, Clanwilliam, Wellington, Porterville, Caledon and Malmesbury.

The main reasons why the remaining centres in the Union have not yet adopted the composting scheme are briefly the following:

1. *Lack of sufficient capital to construct the necessary pits.* The cost of constructing such pits varies from place to place, depending on the cost of material and labour, but anything from £15 to £20 per pit can be taken as an average. Villages and some of the small towns, looking at the matter more from a financial point of view, felt that the output might be so small that it would not warrant the expense.

2. *Lack of sufficient quantities of raw materials, especially dry refuse, to absorb the liquids contained in the night soil.* In some parts of the country, where the rainfall is low and poorly distributed, the vegetation is naturally scanty. This creates a real problem which cannot be disregarded. At the same time, the climate and type of farming in these areas are such that there is hardly a demand for compost, which means that this product would have to be exported to distant localities, thus raising the cost and leaving only a very small margin of profit, if any at all.

3. *The decision of the Department of Labour that urban composting schemes should fall under the Factory Act.* The application of this Act meant that the provisions of certain clauses applicable to modern, well-equipped factories had to be complied with. Although it was added in the proclamation that exemptions in certain respects could be granted, many centres did not see their way clear to adopt the process under such conditions.

4. *Uncertainty in regard to the demand for the final product.* This question was asked in practically every instance and the fact that the Department was not prepared to guarantee either a price or a constant demand for the product made the scheme less attractive. There is, of course, always the possibility that the demand may decline after the war when supplies of artificial fertilizers will again be available. It is nevertheless felt that as the supplies of Karoo manure are being exhausted

since the restriction of the importation of artificials, compost may take its place as a worthy substitute.

5. *The mercenary attitude of many local bodies.* In many cases town councillors were interested in the project only because they regarded it as a potential gold mine. When it was explained to them that they should at most hope for an appreciable reduction on the cost of night soil and garbage disposal, the scheme lost its attractiveness. Many of the municipal compost works are charging excessive prices in an endeavour to show clear profits. In their balance-sheets the costs of disposal under the old system are usually ignored and the national service that is being rendered by making compost is entirely lost sight of.

Whatever the arguments are, one is forced to the conclusion that finance is the major consideration and that unless the venture can be proved to be a sound financial undertaking all the advantages attached to the adoption of such a process, from a sanitary, hygienic, anti-waste, or health point of view, seem to count for very little. Fortunately there are exceptions where urban authorities look upon the composting process as something that has come to stay whether the demand for the product remains what it is to-day or not. In this they find a substitute for a costly sewage scheme, for which they may never hope to raise enough funds. Many of them have already come to the conclusion that most of their disposal problems can be solved in a sanitary, hygienic, and profitable way by the adoption of the urban composting process, provided it is carried out under properly trained supervision.

6. *Lack of interest.* This was found to be due either to ignorance or wrong interpretation. In coastal towns and cities sanitary disposal problems are "solved" by way of dumping the material recklessly into the sea. To a certain extent, however, an exception was found in the case of Durban, one of the biggest coastal centres. Here the Director of Parks and Gardens has set a worthy example to other similar centres by producing about 1,000 tons of compost annually from organic refuse on the true Indore principle, instead of allowing such materials to be passed through the city's incinerators.

The same lack of interest was encountered in large inland centres with properly equipped sewage disposal schemes. Their objections were in many instances well grounded as the adoption of a composting scheme would have meant the carting of raw materials over considerable distances to the site of the actual disposal works, thus making the scheme not only unpractical but also uneconomical. Fortunately in such centres compost is nevertheless made according to the true Indore method by Directors of Parks and Gardens, but usually on a scale only large enough for their own demands. The rest of the valuable refuse constituent usually finds its way to incinerators where it disappears in smoke instead of being conserved and used on the land.

7. *Fear of disease dissemination.* In certain areas, especially the subtropical parts of Natal, local as well as medical authorities were afraid that amoebic dysentery might be spread by the use of the final product as a fertilizer. The Union Department of Public Health, however, expressed itself quite definitely on this point by issuing the following statement at the time: "There is no likelihood of the matured compost, used as a fertilizer, acting as a medium for the dissemination of infective material of amoebic dysentery and parasitic worms, provided the process of composting has been carried out in accordance with the instructions issued by the Department of Agriculture and Forestry, where temperatures of 150° to 160° F. are attained in the pits for two to three weeks." Although this statement, issued by responsible authorities, sounded convincing enough to most urban bodies, some diehards were nevertheless still encountered. The irony of it all is that some of these very same ardent objectors and critics will no doubt cheerfully buy and eat, without any objection or discrimination, vegetables raised by Indians in the sub-tropical parts on soils fertilized with crude and most probably amoebic dysentery-infested night soil.

Nothwithstanding all these objections and difficulties, which naturally had a hampering effect on a more general adoption of the composting process, the results after two years from the inauguration of the scheme are spectacular and encouraging. From the table given it will be seen that before long this country may have at least 100 urban areas in which this process has been adopted. Although actually only about one-third of the urban centres in the Union are actively engaged in this work, the figures rather tend to give a wrong impression of the true position, since about two-thirds of the total urban population are included in the 100 centres mentioned.

Were it not for the instructional short courses held mainly at Ficksburg, Potchefstroom, Walmer, Fort Beaufort, and Graaff-Reinet, it is doubtful whether the actual position would have been as it is to-day. At these centres the various urban representatives became acquainted with the process in general very much more readily and thoroughly than

282

would have been the case if they had had to be taught by their own experience.

Apart from the above, a very encouraging development has taken place at Darnall in the sugar belt of Natal, where Mr. G. C. Dymond has demonstrated so clearly that the vast quantities of sugar waste could be composted with little difficulty and at small expense to serve the essential purpose of linking up the productivity of soils of the sugar belt with the most important factor in the production of cane or any other crop, namely, yield. The same investigator hopes to prove that it may be possible to prevent the degeneration of varieties by practising such conservation methods. For many years these mountains of valuable sugar waste were burnt or neglected, or their value as a compost manure overlooked, but now many scientists and planters in the sugar industry have become compost-minded. The author was invited to read a paper on this subject at their recent conference held in Durban in April 1944.

Although the practice of burning trash before the cane is cut and transported to the mills may result in a saving of labour and expense, it is nevertheless an extremely wasteful procedure. The sooner some other and less wasteful method is discovered by means of which the plant could be stripped of its leaves in an economical and practical way, the better for the industry as a whole. By virtue of its high organic matter content cane trash is a very valuable fertilizer material when composted with nitrogenous substances. Even though the resultant manure may not be required on the plantation itself (which in itself is still a debatable question), together with megasse and filter press cake it may form a valuable by-product for any sugar concern if turned into compost and disposed of to fruit or vegetable farmers in the vicinity.

Compost is also manufactured at Durban on the Earpe-Thomas principle, mainly from vegetable leaves, fruit peels, leaves, and similar materials. It is claimed that according to this method the composting process can be completed within thirty-six hours by the inoculation of the material with special bacteria. The cost of production of this type of compost, called *Organo,* is very high in comparison with that of urban compost, but chemically there is very little difference between the two. A considerable quantity of otherwise wasted organic material is thus finding its way back to the soil, which otherwise would not have been the case.

In addition, some of the larger inland centres are making available considerable quantities of sewage sludge to market gardeners in their vicinity, while the effluent from sewage disposal works is often used for irrigating artificial pastures.

The above is a brief summary of the position as it presents itself to-day in this country. Very much more could undoubtedly still be done in utilizing the enormous quantities of valuable organic materials which are accumulating daily somewhere within the boundaries of urban areas.

As regards the return of the bulk of such materials to the land in the form of properly prepared compost, the question arises whether the State should not step in and either compel the local authority to make compost under supervision or itself undertake the composting of urban refuse material.

Before proceeding to a brief description of some of the experiments carried out at Ficksburg during the past two years in connection with urban compost, the author would like to give the chemical analysis of some samples, calculated on a dry basis, in the following table:

Origin of Sample	Loss on Ignition	Percentage		
		N	P_2O_5	K_2O
Ficksburg	34.43	1.14	1.42	1.24
Ficksburg	44.94	1.18	0.99	1.46
Ficksburg	39.17	1.12	1.41	1.39
Ficksburg	46.24	1.36	1.34	1.00
Ficksburg	47.61	1.53	1.92	1.08
Ficksburg	49.79	1.40	1.59	1.31
Walmer	44.37	1.54	2.76	1.10
Volksrust	30.21	0.78	1.49	1.11
Alberton	43.64	1.62	1.46	1.93
Bethlehem	42.30	1.58	0.90	1.19
Bethlehem	38.03	1.41	1.11	1.31

According to these figures and other observations made at Ficksburg, urban areas in the Union of South Africa are annually accumulating: 230 to 240 thousand tons of organic matter; 15.7 to 26.2 million pounds of nitrogen; 5.4 to 9.3 million pounds of potash; and 5.2 to 8.8 million pounds of phosphoric oxide, in the form of human excreta and town refuse. (The urban population is taken at about 3.3 millions for Europeans and non-Europeans.) Of these quantities, at least 50 per cent is lost or destroyed in one way or another with the result that, no matter how thorough the methods of salvaging and conservation, the quantity ultimately returned is only about half. The longer the return of this material to the land is delayed, the greater is the actual loss. The composting of urban refuse, therefore, is not only an essential but also a most urgent duty resting on the shoulders of those responsible.

As far as the process itself is concerned, in any composting scheme there is one dominating factor which must be borne in mind continually and that is *temperature*. This factor is not only an indication of the success with which the process is being carried out, but also determines the degree to which fly maggots and harmful pathogens may be destroyed. Temperature, therefore, may serve as one of the best indications of the success of a composting process. If it fails to develop, everything goes wrong: if, on the other hand, it develops favourably, we may take it for granted that the process is being carried out properly and successfully.

In the experiments carried out at Ficksburg since the inauguration of the national scheme temperature, therefore, played a major role. In view of the fact that harmful pathogens are destroyed at certain temperature levels if subjected to such temperatures for varying lengths of time, an experiment was carried out to determine average temperature ranges in an urban compost pit, the results being as follows:

Temperature Range in Degrees F.	Time Expressed as Percentage of Total (30 days) at which Compost Material was Subjected to such Temperature Range
51— 60	1.78
61— 70	1.11
71— 80	2.89
81— 90	2.00
91—100	1.77
101—110	2.22
111—120	2.67
121—130	9.55
131—140	24.67
141—150	33.33
151—160	18.00

It may at the same time be stated that this experiment was carried out during the winter months when the minimum temperatures were as low as 18° F.

If 125° F. could be regarded as the minimum safety limit (cysts of amoebic dysentery, for example, are destroyed at 122° F. in two minutes) then one may conclude that the material in a compost pit is exposed to temperatures above this limit for 80 per cent of the time and that the possibility is, therefore, exceedingly small of harmful pathogens surviving or being disseminated when subjected to such limits of heat over such long periods.

Temperature has also an important bearing on the extent to which flies will breed in a compost pit. Flies are not only a nuisance but a menace, since they are largely responsible for the spread of certain diseases and epidemics. After a careful study the conclusion was reached that, wherever excessive numbers of flies are encountered at a compost site, this may be taken as an indication that the process is not going properly, the most probable cause being carelessness. Experiments conducted at Ficksburg in this connection have proved that 85 per cent of the maggots present in the compost material during the process can be destroyed by giving the contents a thorough turning. The heat generated as a result of this will be sufficient to destroy them, provided the material containing such maggots is buried in the centre of the pit where, as a rule, the temperature is very much higher than at the bottom or along the sides. Naturally it is impossible to kill all the maggots in this way and some of them will ultimately escape as full-grown flies, but if poisoned bait is put out these may be got rid of as well. In the early stages of the process fly maggots fulfil a rather important duty, since they help to break up lumpy material, thus bringing about better aeration and advancing the process in general. They should, however, be carefully watched and destroyed as soon as their job is done, otherwise they may complete their life cycle and cause endless trouble.

During periods of excessive rain one cannot rely on the above method alone, namely, that of killing maggots by working over the contents of pits, as the rain tends to cool down the material before the maggots are destroyed. An experiment was therefore conducted with certain chemicals harmless to the process but harmful to the maggots. Two relatively cheap by-products of the Iscor Steel Works were tried out. These were crude naphthalene and interstill residue. The former was used in a fifty-fifty mixture with sand, scattered over the surface of material and lightly worked in, while the latter, emulsified with soap water and used in a 4 per cent strength, was sprayed over the surface of the material in a pit. Both of these chemicals proved effective enough to destroy about 80 to 90 per cent of the maggots during excessively wet periods, when ordinary turning of the contents could not be resorted to. At the same time, these chemicals appeared to have no ill effects on the development of the process itself, judging by temperature observations during the experiment.

For the above two reasons alone it ought to be the aim of every compost producer to obtain as high temperatures as possible in the compost pits under his supervision. There are certain external influences, how-ever, over which one unfortunately has no control. Such factors are rain and atmospheric temperatures.

During the coldest months of the year a rainfall of 1.5 to 2.5 inches had the effect of decreasing the temperature in a compost pit by any-thing up to 15° F. On the other hand, a fall of less than 1.5 inches had no material effect on the temperature in a compost pit at all, and obser-vations seemed to point to the fact that such precipitations may be expected to promote rather than hamper the process.

Minimum atmospheric temperatures of 16° F. to 18° F. during the winter months caused the temperature in a compost pit to drop only 2° F. This only seems to happen when temperatures fall to 20° F. or lower; above this, it was shown to have no material effect at all on compost temperatures.

Factors which influence the temperature in compost pits and over which definite control can be kept are depth of pit and quantity of night soil added per volume of dry refuse. Experiments proved that a four-foot depth of pit gave rise to about 30 per cent higher temperatures than did a two-foot pit, while a proportion of one gallon of night soil to one cubic foot of dry refuse gave the best result as far as temperatures were concerned. The wider the ratio of the latter, the slower the rise in temperature and naturally the longer the time before the process is completed.

Until such time as further tests are carried out, it may be stated that preliminary experiments seem to indicate that during the ripening process, over a period of six months, urban compost did not undergo any material change chemically, whether stored in the open or under protection. A reduction in volume may, however, have taken place in the meantime.

Urban compost production in South Africa has undoubtedly come to stay. To most of the municipalities in the country who have adopted the process this way of refuse disposal means more than just a possible source of extra revenue or an answer to the call of the Department of Agriculture and Forestry to produce compost in order to relieve the fertilizer shortage in the country. To such centres it means, in the majority of instances, a solution to long-standing sanitary and other disposal problems that called for urgent attention long ago. It offers above all a hygienic, harmless, beneficial, and economic method of dis-posal of obnoxious collections accumulating in urban areas, where up to now these valuable, though dangerous, materials were merely lying scattered or buried on town commonages, as a constant source of nuisance and possible disease infection. Furthermore, a proper composting process renders such materials harmless in a quick and efficient way, and may ultimately result in creating a healthier environment for congested communities.

Cities and towns have for too many centuries been veritable graveyards where, in most instances, only the charred remains of the youth and life of many a soil—and ultimate civilization—lie buried and forgotten. It is our duty, as well as our privilege, to ensure that such destructive, almost criminal, practices are no longer allowed to continue. It is sincerely hoped, therefore, that this brief description of what has been done along these lines in South Africa will serve the worthy purpose which Sir Albert Howard intended when he sent me his kind invitation to write this appendix.

If we are "to endure, if we are to project our history, through four or five thousand years, as the Mongolian nations have done," according to the late Professor King in *Farmers of Forty Centuries*, "we must re-orient ourselves; we must square our practices with a conservation of resources, which can make endurance possible."

Ficksburg,
 South Africa.
  19th December 1944.

APPENDIX D

FARMING FOR PROFIT ON A 750-ACRE FARM IN WILTSHIRE WITH ORGANIC MANURES AS THE SOLE MEDIUM OF RE-FERTILIZATION

BY FRIEND SYKES

THE TASK of compressing into an article of 4,000 words and yet doing justice to the story of the enterprise indicated above is no easy undertaking. The whole story needs the book now in course of preparation which is likely to be published by Messrs. Faber and Faber in due course.

For the last hundred years neither farming nor farmers have received at the hands of their fellow citizens a "fair crack of the whip." With ideas on trade and international commerce founded upon a thesis which has proved to be without equal in unsound thinking, with conceptions of economic theories which are as far apart from true economics as the North Pole from the South, our industrialists and their political counterparts have, since the year 1846 which saw the passing of Peel's Corn Laws, sold the farming of England for industrial gain. Slump has succeeded slump, unemployment has become an incurable cancer in our lives, upon one great war has followed a still greater war within the space of twenty years, all showing that something somewhere is wrong with our way of life.

Few industrialists, viewing their declining exports, would ever think that the cause of this vanishing trade was brought about by their own neglect of the agriculture of their native land. They would, indeed, be surprised should this even be suggested to them. But such, nevertheless, is the case. They have built up a false doctrine that without exports this small island of Britain simply cannot live. They are without any panacea for re-establishing that trade, because they, too, recognize that the countries which were their one-time customers are now not only making for themselves the goods they once bought from us, but because of even better methods than we were wont to employ can now beat us in open competition in those few remaining world markets which are, though in diminishing quantities, still buying goods from outside. So that the further we go, the more complex and insoluble becomes the

289

economic problem which this country—and the universe—has got to face.

In what way can agriculture contribute towards bringing order out of all this chaos? Can cosmos emerge out of chaos? Yes, definitely. Agriculture is the fundamental industry of the world and must be allowed to occupy a number-one position in the economy of all countries. The story of Chantry Farm, Chute, Wiltshire, points the way.

We must begin by making one basic assumption: That a farm is analogous to a country and in matters of foodstuffs it must sooner or later become self-supporting. Like a country, again, it cannot entirely ignore trading with the outside world, for the farm requires tractors and implements, buildings and other things, which it cannot provide for itself. *Food, however, must be produced at home,* and any produce in excess of that required for the farm's own human population and its livestock can be sold *in exchange* for those implements and services which are the production of citizens not engaged in farming. The farm and the country, therefore, are in every respect analogous, and this simile must be borne in mind, firstly in order clearly to understand the message implicit in this farming story, and secondly in perceiving the practical application of this lesson to the rectification of the ills of the world which are entirely man made.

After having farmed in Buckinghamshire and elsewhere for over twenty years, I eventually migrated at the age of forty-eight years to an estate of 750 acres on some of the highest land in Wiltshire. This property lies on the eastern escarpment of Salisbury Plain. It is situated in the parish of Chute and at its highest point lies some 829 feet above sea level. It is windswept and bleak. These features are somewhat redeemed by a southern aspect, but, on the other hand, are counter-balanced by the force of uninterrupted gales from the south-west whenever the wind comes from that direction. The land was more or less derelict, and in the records of title which I examined I found that a very large number of so-called farmers had occupied this plot of earth in the course of some sixty years, each of whom had been forced to leave the bleak, unprofitable farm because they were financially worse for wear, or likely to reach insolvency if they continued in occupation. The whole estate was exposed for sale in 1929 and at 50s. per acre freehold it could find no purchaser. It was just "space out of doors," as one of my farmer friends described it, "and not fit for any decent farmer to occupy."

There is evidence in the ancient barrows to be found on the property that this piece of agricultural land has its farming roots embedded in remote antiquity. We have had incidents of discoveries from time to time which show that history has been written here before, both in farming lore and in "bloody battle," for here was fought the Battle of the Bloody Fields some four thousand years ago. This land was probably among the very first that the earliest inhabitants of these islands attempted to cultivate and live upon, land such as Sir Albert Howard had in mind when he wondered "whether there ever would arise a farmer in our own time who would attempt to wrest a living from the highlands of our chalk country and cultivate again the lands which were the first to be farmed in England and which, because of their poor quality, their remoteness from towns and railways, and their altitude and other disadvantages" had been lost to British agriculture. Visiting Chantry for the first time a few years ago, he uttered an exclamation of delight that at long last this dream of his had really come true, for here he saw this ancient piece of England under the plough and in course of re-fertilization according to the rules of good husbandry, as we understood the meaning of that term in the days of our great-grandfathers.

A quite reasonable query may here be asked: If the story of this farmer is worth even the reading, to say nothing of the writing, why should he, if he knows anything about his job, deliberately take a piece of waste land possessed of these obvious disadvantages? Surely, if he has indeed the knowledge of farming which the writing of this chapter suggests must be his, he could have found a more useful sphere in which to expend his time and talent, and withal make "more out of much instead of making a little out of next to nothing." And I entirely agree, but when I took on these obvious difficulties and obligations I did so with my eyes wide open. As a land valuer I have had no little experience; I have surveyed and valued land in nearly every county in Britain from Aberdeen to Cornwall. I have seen farming throughout Britain in many phases of its practice. Few people could have been more conscious of the magnitude of the task that I voluntarily imposed upon myself when, in 1936, I came to live upon and farm these now most beloved, but then forlorn and derelict, acres.

The whole question depends upon what object you are pursuing when you begin any task that really matters, and one of the lessons of my experience and observation of farming everywhere was that livestock are inseparable from good farming, that the best and most stalwart of all stock appeared to be produced on the highlands, that hardy climatic

291

conditions were the invariable accompaniment of constitution and health in livestock, and, moreover, the saying of that old septuagenarian Wensleydale farmer in Muker market place still rings like a clarion call in my ears, "Remember, young man, the higher the land, the sweeter the herbage, the better the cheese." This is no mere tale told for the sake of humour, and those who have the inherited attribute of "farming-in-their-bones" will feel that instinctive respect for those country sayings, which are usually founded upon the kind of wisdom which has close observation of Nature as its university. Furthermore, we are breeders of racehorses and I have found that the best thoroughbreds are all bred on land with high lime content—either limestone or chalk. Here, in this otherwise wasted "space out of doors," I saw the raw material out of which I could breed and develop bone of that density and texture which is only to be found in the cannon-bone of the deer, and where constitution and stamina would be outstanding characteristics. When the reader appreciates this, he will understand that there was some method in my madness in taking on the burdensome responsibilities which the reclamation of this large farm involved.

The farm from which I came in Buckinghamshire, Richings Park—Rich-ings means rich meadows—was in a belt of the richest land to be found in these islands. One hundred pounds per acre was paid readily by buyers—a striking contrast to the land I was to take at Chantry. Richings, however, from my point of view had severe limitations. For the growing of market-garden crops it was almost unequalled, but the bone in both cattle and horses did not develop well or soundly. My observations throughout England had taught me that the vales and the rich lands were useful to fatten a bullock, but were not the place to breed him. When this fundamental truth is fully appreciated in Whitehall, we may one day have an agricultural policy of greater enlightenment than any ruling to-day, a policy which deliberately fosters and encourages stock breeding by every means, using our hill farms for this purpose and leaving the lower-lying farms for the finishing of those hardy "stores" which the hills have bred. That this is an unassailable fact I have proved to my utmost satisfaction. The hills breed constitution, bone, stamina: the vales develop the fat. Our agricultural livestock policy, therefore, should visualize the hill farm as the true complement of the farm in the vale.

Before we came to Chantry I proved my theories in this regard at Aston Tirrold in Berkshire, where for years I kept thoroughbred mares.

Here I had the good fortune to breed Statesman, by Blandford ex Dail, who ran third in Hyperion's Derby and was the winner of several important races; he is now the leading stallion in Japan and the sire of one of Japan's Derby winners. Another high-class animal we bred was His Reverence, by Duncan Gray ex Reverentia; this horse won ten prominent races with a total of over £8,000 in stake money. Solicitor General, a good racehorse and now among the élite of New Zealand sires, was another animal bred on the chalk hills above Aston Tirrold.

To-day at Chantry there stand seven distinguished thoroughbred mares, with foals at foot and with yearlings in the other paddocks, of a class and quality better than any we have ever bred. These achievements, regarded by many people as rather outstanding, are the result of the work we carried out at Chantry in bringing this derelict countryside into a system of agricultural usefulness, where this land now vies with the best in England for the weight of its crops, their health, and their general excellence. Horses are my life love, and I could indeed fill a book with interesting experiences in connection with their breeding and their subsequent performance; but space herein calls for abbreviation and I must now refer to our cattle.

At Richings Park we first of all bred Friesians. We had the finest foundation stock that could be obtained. Many of those cows were from the original herd which won the Silcock Five Hundred Guinea Cup. We ourselves won the One Hundred Guinea Makbar Gold Cup for the best herd of dairy cattle in the three counties of Oxfordshire, Buckinghamshire, and Berkshire. One of our cows, the famous Kingswood Ceres Daisy, was for several years the European Champion in so far as she gave 6,600 gallons of milk with her first three calves. At the Royal Show our stock was often in the winning lists.

In Berkshire pigs, of which we have been breeders for many years, we won the supreme championship at the Royal Show at Leicester in 1924. Progeny from this sow was exported throughout the world and reference to her was often made in catalogues of pedigree pig sales. She was regarded as the finest example of a pig that had been seen at the Royal Show for forty years. The list of our winnings shows interesting achievements, but all this success was to receive a severe check one day.

"Vicissitudes of fortune, which spares neither man nor the proudest of his works, which buries empires and cities in a common grave."

And, indeed, so it happened to my two brothers and to me, for our long run of achievement in livestock production was to end with dramatic suddenness. The Ministry of Agriculture had been made aware by medical and public opinion that all was not well with the nation's milk supply and by way of grading up the dairy cattle the first Accredited Milk Scheme was inaugurated. As one of the leading breeders, we were asked by the University of Reading to show the way to other livestock men by submitting our herd to the Tuberculin Test. We agreed. Judge of our surprise when 66 per cent reacted—the premier herd of the three counties—what must have been the condition of the other dairy herds in that area?

This startling result gave us much food for thought and it was some time before we could diagnose the cause. We pedigree breeders have a saying: "50 per cent of the pedigree goes in at the mouth." Therefore we concluded there must be something amiss with our system of feeding, and we eventually suspected that the cow with her four stomachs was not a concentrated food converter, but, in her natural surroundings, a consumer of roughage. Were not the highly concentrated cakes with their well-known stimulating abilities for the production of rivers of milk the cause of the decline of the health and stamina of our cattle? We thought it over. We consulted authorities famous for their eminence. We had produced fantastic milk records, had been accorded the highest awards in the show-rings, but it was at the expense of the health and constitution of the cows.

We then took a decision requiring both courage and action. We would completely reverse our milk production policy; we would feed the cows more normally, abandon high milk yields, and make the health and constitution of the cattle our primary object and milk production secondary. We held a dispersal sale of our valuable Friesian cattle which had taken so many years to breed and which had, in the eyes of the showman and record-breaker, achieved so much. We then went in for Channel Island cattle, and here good fortune again attended us in the show-ring, for we bought as a calf the bull, Christmas of Maple Lodge, which won the supreme championship at the Royal Show at Chelmsford.

But troubles seldom come singly.

> In trouble to be troubl'd
> Is to have your troubles doubl'd.

And at this same period our most valuable thoroughbred mare contracted the dreaded disease, contagious abortion. An eminent veterinarian advised her destruction. I declined the advice and determined on

294

a treatment of my own, which was to turn the mare out into a large paddock where no horse stock had been grazed, where artificial manures had never been used, and where she was condemned to live for two years eating practically nothing but grass. At the end of this period she was examined by a competent veterinary surgeon and declared clean. She was mated by natural means, proved to be in foal, and subsequently bred over the next seven years four valuable colts, she herself living in good health to the ripe old age of twenty-one. Here was my first attempt to cure an allegedly incurable disease by giving the creature nothing but grass grown on land where no artificial manures had ever been applied—in other words, Natures' food from humus-filled land.

In the early nineteen twenties I had the good fortune to meet the late Major Morris of Aston Tirrold, Berkshire. He became the trainer of my thoroughbreds and in succeeding years I was to see and learn much that was to shape my future agricultural policy and practice. Morris was a man of the highest character, education, and farming knowledge. He was years ahead of his time as a grass-grower, and knew how to establish the sward for a racehorse paddock such as none of his generation ever created. His experience was not available to all, but, being both a patron and a friend, I was privileged to learn much from him. From Morris I learned those elementary lessons which stood me in good stead in later years. Morris farmed some 2,000 acres of Berkshire light downland, yet on that thin soil he grew the heaviest crops of grass and clovers I had ever seen.

As the system of farming at Chantry is now regarded as somewhat original, I will detail the plan of management which I formulated when we left Richings with its accumulated experience and began on this very different, light, high-lying land in the mid-western region.

Travelling about England in pursuit of my professional activities in land survey, I had seen widely varying results everywhere and, after over twenty years of actual farming myself and experience obtained from the examination of other people's work, I had got down to a few principles of my own which might here for the first time be stated.

Fertility on all land can be brought about by following four items of farming husbandry: 1. Good cultivation.
                                        2. Clean land.
                                        3. Subsoiling.
                                        4. Organic manuring.

What a volume of literary work these four headings could provide!

Take good cultivation—if there is one craft which the modern farmer has almost completely forgotten (or would it perhaps be truer to say, never learned) it is that of cultivation. Neither in theory nor in practice does one farmer in a hundred realize how important it is to cultivate, cultivate, and cultivate. The old Wiltshire saying, "A season's fallow with good cultivation is worth more than a coat of dung," is of all good old adages the most forceful. If I can lay claim to be a good farmer, or better still if those who follow after me will but say, "He was a good farmer," then indeed my bones will rest in peace; but if 1 have any justifiable claim to being called a good farmer, it is because I believe I really understand, perhaps better than most, the art of thorough cultivation. What exactly do I mean by thorough cultivation?

Let us assume that I am beginning work on a piece of derelict downland, of which I had hundreds of acres when we started at Chantry. My first act of husbandry is to plough that ground *four* inches deep in October with an eleven-inch furrow; this would lie all the winter and have the benefit of rain, snow, and frost; as soon as possible in the spring it would be cross-ploughed; if the weather was favourable and dry, it would be ploughed again in three or four weeks; it would be ploughed again in a further four weeks—four ploughings in all. Then throughout the summer, as often as I could do it, I should cultivate with a Ransomes Equitine cultivator, certainly the finest implement yet invented for doing a really good job of cultivation. I have cultivated four and even six times in the course of a summer. By this means all weed seeds are encouraged to germinate and are ploughed or cultivated back into the ground. Couch, creeping thistle, buttercup, ragwort, and other noxious weeds are killed outright. The land is oxidized so thoroughly that wireworms and leather jackets and all anaerobic bacteria, which cannot thrive in the presence of air, are killed, and the earthworms, fungi, moulds, and microbes—the unpaid labour force of the farmer, there awaiting in millions to serve him as nothing else can if only he knows how to harness this vast army of workers—are ready to prepare the food materials the crops need. If I could persuade the farmers of England to learn these very elementary and fundamental truths, I would give everything I possess to achieve such an end. Scarcely a farmer anywhere really appreciates these all-important facts. I know, of course, that ploughing may cost £1 sterling per acre at each operation, that four ploughings may cost, therefore, £4; similarly that cultivations may cost from 5s. to 10s. per acre according to the nature of the land operated upon, and that £10 or even £12 per acre may be spent upon such a cleaning fallow: *even so it pays.*

My third item is subsoiling. If you do not know what this means, it would not surprise me for when I ordered such an implement at Chantry the agent who took my order said, "What on earth do you want a tool like that for in this God-forsaken country? My firm has been in business over a hundred years and has never supplied such an implement before." What, then, does the subsoiler do, and why do I use it?

From five to seven inches below the surface there is a hard colloidal pan sometimes quite impenetrable by the roots of plants. This has been accumulating for untold centuries. Break this up by means of the subsoiler to a depth of two feet: moisture will then readily sink to the lower strata; deep-rooting plants will go down through those cracks into regions below in search of minerals and trace elements, which are ofter there in quantity and sometimes not available in the surface soils. While moisture will sink down, so it will rise again by capillary attraction when the hot sun is playing upon the surface soil or stimulating the plants into summer growth, causing increased root activity. The difference between using a subsoiler on almost all lands and not using one is perhaps the most dramatic in all farming operations. I have seen land that would not grow anything come into life and produce a heavy crop purely through the use of the subsoiler. A minimum increase of two sacks of wheat to the acre can be expected, yet it would not surprise me at all if claims of an increase of six to ten sacks were made. An eminent farmer, who saw me use a subsoiler, told me he had improved the output of 5,000 acres of his land by 50 per cent since he used this implement. Until you have seen what the subsoiler can do, its beneficial effects cannot be appreciated. Ransomes C.I.C. subsoiler, however, requires a Caterpillar or Track-layer tractor to pull it. Wheel tractors will not touch it. The cost of the operation varies with the type of land, but on this ground, where serious physical difficulties are encountered in large flints underground, it costs about 25s. per acre. A cut to this depth of two feet is made every four feet all over the field. In this way the entire subsoil is broken into fragments underground. No subsoil comes to the surface. This would be most undesirable; you must keep your subsoil underneath and this implement will not bring it up. If a farmer does not possess a Caterpillar, then he can hire one from his County War Agricultural Executive Committee and perhaps they have a subsoiler as well. As a matter of fact, I do not believe that all

297

the War Agricultural Executive Committees in Great Britain do possess one, but if agitation is sufficient, they will all became enlightened and buy one or two.

Lastly, but by no means least, we come to the all-important subject—this controversial subject—of re-fertilization. Of course I believe in organic manuring and do not use inorganic fertilizers. Is this opinion founded upon experience? Most certainly, and these are my findings. A portion of the land at Chantry would not grow cereal crops at all when I took over the land. None of it would grow any good grass; the herbage was not capable of keeping the cattle alive and we had to purchase outside foods, which cost some £80 a month. To-day, after less than seven years of farming, we are growing some of the biggest crops of wheat and grass that can be found anywhere in England. This has been achieved by following the technique already described and by the exclusive use of organic methods of re-fertilization. Let me say, however, with all the emphasis at my command, that unless a farmer is prepared to cultivate thoroughly, he is wasting time and money in applying manure of any kind to his land. The indispensable forerunner of manuring *must be thorough cultivation and subsoiling*. After that we can talk about applying new fertility to the soil, for it is then in healthy balance and in a condition to receive added humus to restore and maintain—and increase amazingly—the fertility of which almost all land is capable.

The systems of applying organic manure to the land employed at Chantry are many and various. Again, unless I could allocate a very long chapter to this one subject, I could not do full justice to it. I will confine myself, therefore, to outlining broadly two systems of fertility renewal.

The first system is to bail the dairy cattle over a mixed ley. The bail is a movable dairy which travels over the fields and secures an even distribution of dung and urine. As a system it stands alone in economic milk production. It also produces milk of T.T. Attested standard—that is to say, the highest grade. The cattle are controlled by electric fencing, so that their dung and urine are evenly distributed over the field. Dairy cattle are fastidious feeders and are given the cream of the feeding, leaving the folds with much unconsumed food. They are followed by Galloway beef-breeding cattle, who eat everything as it comes and clean the leys right down to the ground. Sheep, too, usually join in this roughage clearing. There is thus left a field covered with the dung and urine of three types of animals. The bacterial life of dung and urine of these varying species is important, for it keeps each class of stock in a balance of health that is truly remarkable. Disease is nearly absent, I believe, from this farm now as a result of this method of field-controlled grazing.

At the end of the grazing period the field is harrowed thoroughly, spreading evenly the pats of manure, and then is rested for a short time, during which rain doubtless falls; sheet-composting of the area takes place; worms in thousands visit the surface and draw down below the dung and waste vegetation, which revivifies the soil by increased bacterial activity and breeds untold millions of protein-consuming fungi, moulds, and microbes, all of which—the farmer's best friends and unpaid labour force—are ready to develop an abundance of those plant foods which will produce another heavy growth of grass and clovers ready for further treatment of a similar kind, or a variation of it.

My leys are usually put down for four years. The first year is all grazing; the second, hay and grazing; the third, hay and grazing; the fourth, grazing until June, after which it is ploughed, fallowed until September and then in that month ploughed again. It is then sown with wheat, and crops up to eighteen sacks per acre may sometimes be expected from this *complete* system of farming technique. After wheat a crop of roots can be taken, followed by oats or barley, after that a fallow clean until the following July when rye may be sown, then the land may be grazed until Christmas, and in the spring undersown with a mixture of grass and legumes somewhat of the following composition and quantities:

	lb.
Cocksfoot	10
Timothy	3
Italian ryegrass	3
Rough-stalked meadow grass	1
Crested dogstail	2
Meadow fescues	2
Common milled sainfoin	10
American sweet clover	4
Hants late-flowering clover	2
S.100 white clover	2
Kidney vetch	2
Burnet	3
Chicory	3
Total	47 lb. to the statute acre

It will be observed by any student of botany that here is a mixture of grass and legumes of unusual character and quantities, but my experi-

299

ence has shown that it pays well to sow fairly heavily to ensure a good take. The deep-rooting legumes like sainfoin, sweet clover, kidney vetch, burnet, and chicory are important for the establishment of a good and continuous sward. The roots penetrate deeply for lime, minerals, and trace elements, making these essential materials available for the shallower-rooting grasses, while the nitrogen fixation effected by the inclusion of sweet clover is, perhaps, the greatest magic of all. Furthermore, all these plants are Nature's own subsoilers, and once they are established in the land the necessity for frequent subsoilings with the subsoiler disappears to a large extent.

Now I come to the second system in the scheme of re-fertilization. This is composting. I am a great believer in composting. My men believe in it too, but mostly when someone else is doing the digging and turning. The digging of farmyard muck out of the heavily trodden stock-yard is the hardest, most soul-destroying, and most disagreeable work on the farm. In the old days cheap slave labour from Ireland used to be hired especially to do this sort of wretched job. The way those Irishmen used to handle muck was to me a marvel at which I shall never cease to wonder. But the enlightened English farm worker will not do it, and I cannot honestly blame him. In this connection I well recall an incident of a year or two ago. My men had been muck shifting for weeks. We had moved some 500 tons. It rained and rained; it looked as though it would never stop raining. Their rubber boots were leaking and were filled with squelching liquid manure. One morning they all came to me in open mutiny. "Look here," the leader said, "if this is the only bloody job there is on this farm, we're going somewhere else to work." I sympathized with them and told them to go to the barn and find work there until the weather improved. Meanwhile I went to my old drawing-board and worked upon the designs of a machine to mechanize the muck heap, a question about which I had been thinking for some time. When this was completed, I took the result to Messrs. Ransomes & Rapier Ltd., the famous crane makers of Ipswich, and asked them to make it for me. They examined my drawing and in due course asked me to call. "You have invented something, Mr. Sykes, in this machine. We think we can help. May we do so, and then let us take out a patent together in connection with it?" This was agreed, and the *Rapier muck shifter* is now on the market and accomplishes by mechanical means the most hated of all farming jobs. Why it has never been invented before I do not understand!

What does this machine do? Plate XIV shows the machine itself, and its practical effects on farming technique are quite revolutionary. For instance, we moved an estimated 400 tons of muck, carted, and spread it for £32—a cost of 1s. 8d. per ton. We have never done this by hand for less than 12s. 6d. per ton: 1,000 tons per year is our output of muck. Here is a saving of over £500 per annum, which is more than the cost of the muck shifter.

And here we come to that all-important subject of composting. Composting by hand on the large scale is indeed terrible in both cost and physical fatigue. We can now do the turning mechanically at a cost of a few pence per ton, and we estimate we can turn from 200 to 300 tons per day. Two things, then, are accomplished here: (1) an enormous saving of cost in the preparation of farmyard muck in composting, digging, and loading into carts, and (2) a great saving of time, for we can now do in a few days what previously took months. A fellow farmer said to me one day, "If you stop to tot up the cost of farmyard muck from first to last, you would never put a forkful on to the land; the cost is enormous." With the Rapier muck shifter, however, we have now not only eliminated the high cost of handling and distributing this valuable material back on to the land, but we have so reduced the costs in comparison with *artificials* that economically the latter cannot enter into consideration any more.

The title of this Appendix, "Farming for Profit . . . ," suggests that my final words must relate to the profitable character of farming with organic manures as a whole.

Using all the implements we possess in as effective a way as possible, grazing our cattle by the system of mixed grazing already described, making compost with the Rapier muck shifter, as we now do, I can assure the readers of this book that *organic farming* is not only profitable, but *even more profitable* than farming and re-fertilization with inorganic manures. There are, of course, still further reasons why organic fertilization is better. A healthier livestock is produced. Disease in plants is eliminated. I no longer need to dress my seeds with mercurial dressings. Poison sprays have no place on the farm. The farm is entirely self-supporting. Over 250 head of cattle and sometimes many hundreds of sheep get their living here from food grown on this land. The horses, too, are home supported. We do not find it necessary to change our seeds so frequently, perhaps not at all. We are growing the same wheat here now that we have been growing for six consecutive

301

years. When we bought it—Vilmarin 27—it was subject to black smut. To-day the amount of disease which makes its appearance is negligible. The yields are enormous. And the same results apply to oats and barley.

Lastly, I must refer to *wholemeal* bread. I wonder how many farmers have tried wheat grown with muck or compost as compared with wheat grown entirely with artificials. Not many, I am sure. Then try it. If you have not got any wheat so grown, send for a sack of our wheat and carry out the following instructions. Grind it, just as it is, in the Bamford mill which you doubtless have in your barn. Bake bread from the wholemeal flour so ground according to the recipe to be found in Mrs. Gordon Grant's book, *Your Daily Bread* (Faber and Faber Ltd., London, 1944) and then try your own artificially grown wheat similarly treated, and you will need no further assurance that wheat grown with muck or compost is sweeter to eat, more enjoyable, and more sustaining than wheat grown with the aid of inorganic fertilizers. The incidents I could further relate, showing the increased food value of organically fertilized crops, are many—too numerous to fall within the scope of this appendix.

In conclusion may I express the hope that agriculture may take its proper place in the world of to-morrow and that a public health system may be founded in the future which will be based upon a soil in good heart—a soil that will produce life-sustaining food for both man and beast, which means a soil that is living in every sense of the word. "A fertile soil is one rich in humus" (Sir Albert Howard). "Humus is the product of living matter, and the source of it" (A. Thaer).

Chantry,
  Chute, Andover.
    6th July 1944.

INDEX

A

Acharaya, Dr. C. N., 236
Afghanistan, 6, 131, 133, 134, 176
Africa, 36, 41, 48, 49, 57, 59, 61, 87, 88, 89, 109, 110, 114, 116, 119, 120, 125, 126, 168, 210, 212, 223, 224–234, 274ff.
Agricultural Research Council, The, 80
Agricultural Research Institute (Pusa), The, 3, 129
Agriculture, tropical, 1, 97; Indian, 3–8; primitive, 33–42; on cultivation, 35, 36; Egyptian, 36; Chinese, 38; Greek and Roman, 39, 40, 41; of the Middle Ages, 41, 42, 45, 46, 49; of the Romans in Britain, 43, 44; the two-and-three-field system, 45, 46; Coke of Holkham's development of, 52, 56; extensive, 57ff.; the machine and, 58ff.; the chemistry of, 69; the Ministry of, 76, 248, 250, 253, 260; on research, 77–80; statistics of, 77–80; the Cambridge School of, 248; the Missouri College of, 241
Airy, Rev. W. S., quoted on school gardening, 182
Albrecht, William A., 241
Algeria, 136
Alkali lands, 94–102; the reclamation of, 100–102
America, South, 36, 37, 59, 155
America, U. S. of, 35, 57ff., 60, 64, 65, 87, 88; and the Indore Process, 241–244
Ammonia, 74ff., 189, 205
Amos, Alfred, 2
An Agricultural Testament, cited, 11, 12, 24, 105, 128, 139, 141, 142, 176, 199, 211, 213, 222, 223, 224, 225, 230, 232, 235, 237
Animals, the care of, 4, 5; diseases of, 4, 5, 55, 64, 158ff.; forest, 26, 27; humus and, 31, 32; plough, 34; in the Middle Ages, 41, 42; discarded, 61; the manure (wastes) of, 110, 196, 197, 200, 204–208, 213, 229, 239; pedigree, 163, 292ff.
Ants, 28
Argentine, the, 55
Armstrong, H. E., 78, 248
Ashby, T. W. M., 238
Asia, the dying lakes of, 95, 96, 99; the vine in, 132ff.
Assam, 59, 110, 113, 129, 247
Australia, 57, 59, 62, 87

B

Bacon, Sir Francis, 69
Bakewell, Robert, 51
Baldwin, Stanley, 180
Balfour, Lady Eve, *The Living Soil*, 24, 164, 165, 167, 176, 184
Baluchistan, 5, 132, 133, 134, 142
Banana, the, 141, 142
Barbados, the Imperial Department of Agriculture at, 1; the principal crops in, 1; the sugar industry in, 253
Barley, 34, 51
Barrackpore, 129
Basic slag, 198, 199, 200, 201, 202
Basutoland, 89
Belgic tribes, the, 44
Benares, 108
Berry, Sir Walter, 2
Beveridge, Sir William, investigation of wheat yield by, 49
Black Death, the, 49ff.
Blackheath, 139, 144
Bledisloe, Viscount, 116, 181
Bog (and morass), 27, 28
Borsodi, Ralph, 243
Botany, 18
Branch, Rev. G. W., 118

Bread, 55, 173
Britain, historical survey of agriculture in, 43–56; the submarine menace to food of, 76; and the Indore Process, 238–240; blunders of, 75, 76, 260
British Mycological Society, *Transaction of*, cited, 126
Broadbent, H. R., quoted on devastation of the soil, 64; and the Broadbalk Trials, 75
Bromfield, Louis, 243, 244
Bunting, A. H., 250, 251, 252
Butler, 126
Buy an Acre, Paul Corey, 242

C

Cabbage, the, 24
Cacao, 110, 118–121; estates, 119, 120
Calcutta, 143, 247
California, 102
Cambridge School of Agriculture, the, 248
Cambridgeshire, 45
Canteens, 183, 184
Capetown, 231
Carlyle, Sir Robert, 4, 158
Carpenter, Brodie, 185
Carrel, Alexis, 176
Caterpillars, 107
Caudwell, George, 151
Ceylon, 35, 59, 61, 110, 111, 112, 113, 119, 235, 246, 247
Channel, the, 45
Chapman, Dr. G. B., on produce of humus-filled soil, 184, 185; alluded to, 238
Chemistry, 69, 70, 99, 100, 228, 234
Cheshire, The Local Medical and Panel Committees of, 11, 12, 56, 176, 223, 235, 239, 240
Chicory, 101
China, a waste-utilizing practice of, 7; the agriculture of, 38, 39, 215; the peasant population of, 57; the drainage problem of, 91; alluded to, 205
Cicero, 40
Civil War (English, 1642), the, 51
Civilization, agriculture and, 33; Peruvian, 36; past, 93; failure of, 234ff.
Clarke, George, 5, 105, 137, 229
Clifton Park System, the, 56, 101
Clubs, 4-H, 243
Coal, 53
Cockle, Park Experiments, the, 199
Coffee, 109–111
Coke of Holkham (Norfolk), 52, 56
Cole, Sir Edward Hearle, 101, 102, 125, 223, 237
Coleyana Estate (Punjab) the, 101, 102, 125, 223, 237
Combine, 58ff.
Commons, 46
Compost dressings, Chinese, 8; the Indore Process of (*see*); the *News-Letter* on, 12, 164, 168, 178, 185, 202, 223, 235, 239, 240; for sugar-cane, 108, 109, 224–229; for tea, 111–118; for cacao, 118–120; for cotton, 121–127; for rice, 127–129; for bananas, 141–142; for apples (and other fruits), 144; for peaches, 139–141; for grapes, 132–138; for wheat, 129–132; for tomatoes, 143–144; for strawberries, raspberries, 144–146; for tobacco, 147; for leguminous crops, 148ff.; for potatoes, 150ff.; a quotation on the making of, 153–155; *The Compost Gardener*, cited, 186; sheet-composting, 207, 208, 210; composts described, 211ff.
Connecticut Experiment Station, the, 29
Copra, 235
Corey, Paul (*Buy an Acre*), 242
Corn Laws, 54
Costa Rica, 110, 223